Mastering Fiber Optic Connector Installation

-The Guide To Low Loss, Low Cost, And High Reliability

Version 1.0

Eric R. Pearson, CFOS

Pearson Technologies Inc.

Acworth, GA

Disclaimer

All instructions contained herein are believed to produce the proper results when followed exactly with appropriate equipment. However, these instructions are not guaranteed for all situations.

Notice To The Reader

The publisher does not warrant or guarantee any of the products described herein or perform any independent analysis in connection with any of the product information contained herein. Publisher does not assume, and expressly disclaims, any obligation to obtain and include information other that provided to it by the manufacturer.

The reader is specifically warned to consider and adopt any and all safety precautions that might be indicated by the activities herein and to avoid any and all potential hazards. By following the instructions contained herein, the reader knowingly and willingly assumes all risks in connection with such instructions.

The publisher makes no representation or warranties of any kind, included but not limited to, the warranties of fitness for particular purpose or merchantability, nor are any such representations implied with respect to the material set forth herein, and the publisher takes no responsibility with respect to such material. The publisher shall not be liable for any special, consequential, or exemplary damages resulting, in whole or part, from the readers' use of, or reliance upon, this material.

The procedures provided herein are believed to be accurate and to result in low installation cost and in high reliability. However, there is a possibility that these procedures may be unsuitable for specific products or in specific situations. Because of this possibility, the operator should review, and follow, the instructions provided by product manufacturers. The instructions contained herein are not meant to imply that conflicting instructions from manufacturers are in error.

Trademark Notice

All trademarks are the property of the trademark holder. Trademarks used in the installation documents include Kevlar™ (DuPont), Hytrel™ (DuPont), and ST-compatible (Lucent).

Published by

Pearson Technologies Inc.
4671 Hickory Bend Drive
Acworth, GA 30102
770-490-9991
www.ptnowire.com
fiberguru@ptnowire.com

Version 1.0

Printed in the United States of America.

10 9 8 7 6 5 4 3 2 1

ISBN 978-1466470699

TABLE OF CONTENTS

TABLE OF FIGURES

AUTHOR'S PREFACE

Installing fiber optic connectors is not difficult: when my two sons were 10 and 13, I trained them to install connectors. They achieved the three goals of installation: low power loss, low installation cost, and high reliability. They did so, even though they had their CD players plugged into their ears! (Now I've dated myself!)

This text guides you to achieve these three goals. This is no idle boast: in training installers, I have observed the results and refined these procedures to include only those instructions that work for almost all trainees. The procedures in this manual reflect refinement from 21 years of training, more than 500 presentations, and more than 7900 trainees. With very few than exceptions, all trainees have achieved these three goals! So will you.

This text guides you through successively increased understanding and knowledge, from basic to subtle. Chapter 1 provides the basic understanding of connectors in the network. Chapter 2 provides a detailed understanding of the language of fiber connectors: their functions, structure, performance, types, similarities, advantages, and installation methods. With this understanding, you can understand Chapter 3.

Chapter 3 presents the principles of installation for four commonly used methods. This understanding of the principles underlying the procedures has three benefits. First, you know the consequences of failure to follow the instructions. Second, you are more likely to follow the instructions than you would be without this understanding. Third, you perform troubleshooting with an extensive understanding of the potential causes of high loss and low reliability.

Chapter 4 presents instructions for inspecting connectors that require polishing. These instructions show you how to inspect, rate, and diagnose causes of high loss. With this chapter, you can easily identify causes of high loss and appropriate corrective actions.

Chapters 5-8 present detailed instructions for four commonly used methods. Each set of instructions guides you to achieve the three goals.

These instructions include 'do not do's' and cautions. These 'do not do's' and cautions help you avoid the commonly-made errors I've observed during training of more than 7000 installers. With minor modifications, these chapters can be used to install or train with any connector available.

In addition, each chapter contains two useful sections: a troubleshooting section and a one-page summary. The troubleshooting section helps you recognize the symptom of an error and identify the step(s) at which the error occurred. This section speeds up achieving the three goals. During field installations, you can use a copy of the one page summary as a guide and reminder.

Chapter 5 presents installation and polishing of multimode connectors. In addition, it contains polishing instructions for singlemode epoxy and quick cure adhesive connectors to achieve -50 dB reflectance. Chapter 6 presents installation and polishing of multimode connectors with quick cure adhesive. Chapter 7 presents installation and polishing of multimode connectors with hot melt adhesive. Chapter 8 presents installation of both multimode and singlemode connectors with the no-polish, no adhesive method, also known as the 'cleave and crimp' method.

This manual contains review questions for Chapters 2-8 to assist you and the trainer in assessing and reinforcing understanding. When used prior to hands on training, these questions can lead to excellent results, both in training and in field installation.

For subjects beyond the scope of this manual, this manual contains references to Professional Fiber Optic Installation. Such references appear as: PFOI, section number. For example, (PFOI, 8.3).

This manual is one of a series on Mastering Fiber Optics. Published manuals are:

Professional Fiber Optic Installation-Essentials For Success, described at:
http://www.ptnowire.com/Sfoi-outline-7.2.htm

Mastering The OTDR-Trace Acquisition And Interpretation, described at:

http://www.ptnowire.com/MTO-main-page.htm

Future publications are:

Mastering Fiber Optic Testing

Mastering Fiber Optic Network Design

Best Regards,

Eric R. Pearson, CFOS/T/C/S/I

Pearson Technologies Inc.

fiberguru@ptnowire.com

770-490-9991

http://www.ptnowire.com/

November 2011

1   ESSENTIAL CONCEPTS

Chapter Objective: you learn the basic concepts of fibers and cables.

1.1   PRIME FUNCTION

The prime function of the fiber optic link is minimization of optical power loss between a transmitter and a receiver. This function results from the fact that the fiber optic link must deliver sufficient optical power to the receiver. Without sufficient power, the receiver cannot convert the optical signal to an electrical signal accurately.

In support of this function, the connector provides low power loss. The connector controls power loss through alignment of the fiber to other fibers, to the transmitter, and to the receiver of a link (Figure 1-1).

Figure 1-1: Simple Link

While alignment supports the optical function, it does nothing to provide reliability. The connection of the connector to the cable provides this second function. From this brief introduction, it becomes apparent that the connector installer requires some basic information about both fibers and cables.

1.2   FIBER

The fiber has three regions:

➢   Core

➢   Cladding

➢   Primary coating

1.2.1   CORE

Most of the optical energy travels in the core, the center of the fiber. Two factors result in low loss connections:

➢   Precise core alignment

➢   Matched cores

The connector provides precise alignment. However, without matched cores, even perfect connectors cannot provide low loss. Thus, all fibers in a link must have the same core. Same core has two meanings:

➢   Same multimode core diameter or

➢   Same singlemode mode field diameter

µThe core diameter, in microns (µ), can be large or small. In a data communication fiber, a large core has a diameter of either 50 or 62.5-µ (Figure 1-2). Such a fiber is a graded index (GI) multimode fiber.

In the past, the 62.5µ fiber dominated the multimode data world. In the future, the 50-µ fiber is expected to dominate.

A= core= 50 or 62.5 µm

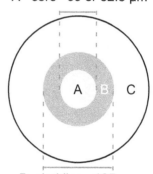

B=cladding= 125 µm
C= primary coating

Figure 1-2: Multimode Fiber

Differences between the chemical compositions of the core and cladding result in a 'critical angle' or angle of acceptance. Rays of light that enter the core within this angle remain in the core and travel along the fiber. This angle defines the numerical aperture, or NA, of the fiber (Figure 1-3).

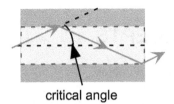

critical angle

Figure 1-3: Fiber Critical Angle

This angle, or NA, becomes important to the connector installer. If the installer bends the fiber, he changes the angle at which light strikes the core boundary. If the bending is excessive, light can escape from the core. In this case, a properly installed connector will exhibit a high loss. A subtle form of this bending can occur under a boot, if the connector boot fits tightly over the cable jacket or buffer tube.

The graded index core has multiple regions of different chemical compositions. These regions enable the fiber to transmit high bandwidths. Differences in these multiple regions result in two types of multimode fibers: standard and laser optimized (LO) fibers.

Standard fibers transmit high bandwidths to distances defined in the data standards. Laser optimized fibers transmit these bandwidths to distances greater than those stated in the data standards.

core≈ 8.2-10 µm

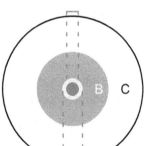

MFD≈ 9.2-11 µm
B=cladding= 125 µm
C= primary coating

Figure 1-4: Singlemode Fiber

In data communication, telephone, and CATV networks, a small core has a diameter of 8-10-µ. Such a fiber is a singlemode fiber (Figure 1-4). In the singlemode fiber, a significant portion of the optical power travels in the cladding.

Because of this fact, the singlemode core diameter is not critical. Instead the singlemode 'mode field' diameter (MFD) is. The MFD is approximately 1-µ larger than the core diameter.

As the core in a singlemode fiber is much smaller than that in a multimode fiber, singlemode connectors are manufactured to tight tolerances. These tolerances make the singlemode connector more expensive than the multimode. Finally, these tighter tolerances make feeding the fiber into a singlemode connector slightly more difficult than into a multimode connector.

1.2.2 CLADDING

The cladding has two functions:

> Confinement

> Increase in size

First, by surrounding the core, the cladding confines the light to the core. This confinement results from the difference in chemical compositions of the core and cladding.

The cladding confines the light by two mechanisms: reflection and wave guiding. In multimode fibers, the light reflects at the core boundary, as long as the angle of incidence is less than the critical angle. The numerical aperture (NA) of the fiber defines this angle. In singlemode fibers, wave guiding confines the light to the core.

The second function is increase in size. This increase results in increased strength and ease of handling.

1.2.3 PRIMARY COATING

The primary coating protects the fiber from mechanical and chemical attack. This protection enables the fiber to retain its intrinsic high strength during cable manufacturing and for a cable lifetime of over 20 years. The connector installer's main concern with this coating its removal during connector installation. The primary coating diameter is usually 250-µ.

1.3   CABLE

1.3.1   ATTACHMENT METHODS

Much of the reliability of an installed connector results from the attachment of the connector to the cable. This attachment is of three types:

➢ To the cladding

➢ To the buffer tube

➢ To cable strength members

All three attachments are made by the same two methods: adhesive and compression. Strength member attachment occurs on one-fiber cables, aka patch cords and jumpers, two-fiber zip cord cables, and break out cables. In these cables, each fiber has its own jacket, under which are the strength members. We refer to such cables as 'jacketed fibers'.

Strength member attachment to a connector is not done with loose tube or premises cable designs.

1.3.2   DESIGNS

Fiber optic cables are of two types:

➢ Tight buffer tube

➢ Loose buffer tube

1.3.2.1   TIGHT BUFFER TUBE

In a tight tube design, the buffer tube is placed over the fiber with no space between the outside of the fiber and inside of the buffer tube (Figure 1-5). This structure allows only one fiber per buffer tube. This buffer tube strengthens the fiber sufficiently so that it can be handled repeatedly without damage. Because of this strengthening, connectors can be installed on tight tubes.

0.9 mm

Figure 1-5: Tight Buffer Tube

A typical tight tube diameter is 900-µ. Some cables have a tight tube diameter of 600-µ. This diameter is matched to the inner diameter of a connector boot.

The tight tube forms the basic structural element in two cable designs:

➢ Premises (Figure 1-6)

➢ Break Out

The premises design (Figure 1-6) is commonly used. The break out design is rarely used.

Figure 1-6: Premises Cable

1.3.2.2   LOOSE BUFFER TUBE

In a loose tube design, the buffer tube is placed over the fiber with space between the outside of the fiber and the inside of the buffer tube (Figure 1-7 and Figure 1-8). This space allows placement of multiple fibers in a single buffer tube. When the buffer tube is removed, the 250-µ coating is exposed. This coating provides strength that is insufficient for repeated handling (Figure 1-8).

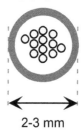

2-3 mm

Figure 1-7: Loose Buffer Tube

As a result, fibers in loose tube cables must be strengthened prior to connector installation. A furcation kit, also known as a 'fan out' kit and 'break out' kit (Figure 1-9) provides this strengthening. The furcation kit includes tubing with a 900-µ diameter.

Figure 1-8: Loose Tube Cable

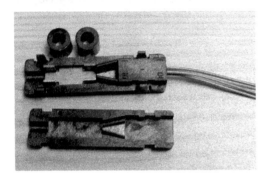

Figure 1-9: Furcation Kit For Loose Tube Cable

1.3.3 BASIC ASSUMPTION

For the purpose of this manual, we assume that the fibers have a 900-µ buffer tube. This diameter results from either a tight tube cable or from a furcation kit on fibers in a loose tube cable.

2 CONNECTORS

Chapter Objectives: you learn the language of fiber optic connectors, the types, their advantages, and performance concerns.

2.1 INTRODUCTION

In this chapter, we present:

- ➤ Function
- ➤ Structure
- ➤ Performance
- ➤ Types
- ➤ Installation methods

2.2 FUNCTION

Connectors provide four functions:

- ➤ Low power loss
- ➤ High fiber retention strength
- ➤ End protection
- ➤ Disconnection

Connectors achieve low power loss through precise alignment of the fiber cores. Connectors achieve high fiber retention strength by gripping the fiber, either through an adhesive or mechanical compression. Connectors achieve end protection by distribution of contact force over an area of increased size, thus reducing the pressure, in psi, imposed on the end of the fiber. Finally, connectors enable disconnection, or detachment, of a fiber from both optoelectronics and other fibers. In short, connectors are used to create temporary connections and are installed in all locations in which permanent connections are not required.

2.3 STRUCTURES

In spite of the apparent complexity and variety of structures, connectors can be described by nine structural characteristics:

- ➤ Ferrules
- ➤ Latching structure
- ➤ Mating structure
- ➤ Key

- ➤ Back shell
- ➤ Strain relief boot
- ➤ Crimp sleeve or ring
- ➤ Simplex or duplex
- ➤ Ferrule cap or cover

2.3.1 FERRULES

2.3.1.1 FERRULE CONTACT

Ferrules in small form factor (SFF) and commonly used connectors make physical contact (Figure 2-1). Most legacy connectors, which are not used in new installations, are non-contact and have losses higher than those of contact connectors.

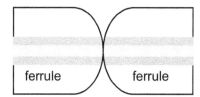

Figure 2-1: Ferrule Contact

This increased loss results from the expansion of the light as it exits a connector. Light exits from a fiber and expands within a cone defined by the NA. In a non-contact connector, the size of the spot on the receiving fiber is larger than the diameter of the receiving core. The light arriving on the cladding of the receiving fiber is lost (Figure 2-2).). A contact connector has no such expansion and, as a result, reduced loss.

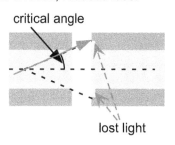

Figure 2-2: Increased Loss From Non-Contact Connectors

2.3.1.2   FERRULE END FACE

Ferrules have three end faces:

> Radius

> Angle physical contact (APC)

> Flat

Contact connectors have a radius end faces (Figure 2-1). Such end faces ensure high performance; that is, low loss and low reflectance (2.4.5). However, this end face cannot provide the lowest possible reflectance. For lowest reflectance, the angle physical contact (APC) end face is required (Figure 2-3). This singlemode connector has a bevel on the tip of the ferrule at 8 degrees to the axis of the fiber. APC products are available for SC, LC, LX.5, and FC connectors.

Figure 2-3: The SC/APC Connector

The APC connectors are not commonly used because of increased cost. Some FTTH networks use APC for low reflectance.

In current generation connectors, flat end faces are not acceptable. Prior to approximately 1985, flat end faces dominated the connector world.

2.3.1.3   FERRULE DIAMETERS

While there are six ferrule diameters, two dominate the industry: 2.5 mm and 1.25 mm. ST™-compatible, SC, and FC connectors have 2.5 mm ferrules. These were introduced in the early to mid-1980s.

LC, LX.5, and MU connectors have 1.25 mm ferrules. These types were first available in the late 1990s. This ferrule has loss lower than that of the 2.5 mm ferrule.

2.3.1.4   FERRULE MATERIALS

The ferrule provides low loss through alignment of fiber cores. With one exception, all connectors have ferrules.

The one exception, the Volition, has no ferrules; instead, it has precision V-grooves that align the fibers (2.5.1.4).

Ferrules can be of different materials. The two most commonly used materials are:

> Ceramic

> Liquid crystal polymer

Occasionally, the ferrule is stainless steel.

As of this writing, ceramic ferrules are required for singlemode connectors, since only ceramic materials can be manufactured at a competitive price and with the precision required for singlemode core alignment.

Unlike the other ferrule materials, ceramic ferrules can be over polished, creating undesirable, excessive undercut (Figure 3-25). Undercutting results in increased loss and reflectance.

The other materials, which are softer than ceramic, can be repolished without expensive diamond polishing films. Diamond films are approximately 30 times the cost of standard films.

This re-polishing removes light to moderate damage, restoring original performance without connector replacement. In addition, soft ferrule materials can be repolished without risk of undercutting. In an apple-to-apple comparison, the soft ferrule materials provide a reduced initial cost and a reduced life cycle cost.

Other connector ferrule materials include:

> Nickel-plated brass

> Aluminum

> Zinc

> Thermoplastic polymer

At the time of this writing, the first three of these materials are rarely used. Thermoset polymer is used in the MTP/MPO (2.5.1.8) and MT-RJ connectors (2.5.1.6).

2.3.2 LATCHING STRUCTURES

Connectors have one of two latching mechanisms:

> ➢ Push on/pull off (PO/PO) latching mechanism
>
> ➢ Rotating retaining ring

All current generation connectors have push on/pull off latching mechanisms. This mechanism has two advantages. First, it makes the connector easy to remove. Second, it enables the connectors to be closely spaced in a patch panel. Close spacing enables doubling and quadrupling the number of connectors in a patch panel. Such close spacing reduces hardware cost.

In the SC connector, the inner and outer housings of create the PO/PO mechanism (Figure 2-4). This structure includes, from left to right: the outer housing; the connector body, which includes the ferrule, inner housing and back shell; the crimp ring; and strain relief boot.

Figure 2-4: SC Connector Design With A Latching Mechanism

Examples of connectors having PO/PO latching are:

> ➢ LC
> ➢ SC
> ➢ MU
> ➢ LX.5
> ➢ Volition
> ➢ Opti-Jack
> ➢ MT-RJ
> ➢ FDDI
> ➢ ESCON

In some connectors, a latching ring, such as that in the ST-compatible connector (Figure 2-5), creates the latching mechanism. The ring must be rotated to latch the connector. The need for rotation increases the space required for each connector.

Figure 2-5: Connector With Latching Ring

This structure includes, from left to right: the connector body, which includes the ferrule, retaining ring and back shell; the crimp ring; and the strain relief boot

Examples of connectors having this type of latching are:

> ➢ FC
> ➢ Biconic
> ➢ 905 and 906 SMA
> ➢ Mini BNC
> ➢ D4

2.3.3 MATING STRUCTURES

There are two structures for connecting, or 'mating' connectors:

> ➢ Two identical connectors or 'plugs'
> ➢ A plug and jack

When the connectors are identical, the connectors require a 'barrel', also known as an adapter, a bulkhead, and a feed through (Figure 2-6). The dominant connector design is of identical plugs.

Figure 2-6: Connectors With Plugs And Barrel

When the connectors are not identical, the connectors form a plug and jack pair (Figure 2-7). The jack performs the functions of a plug and barrel in a single structure. In general, jacks can be flush mounted, while barrels cannot. Flush mounting is an advantage in FTTD networks. Examples of plug and jack connectors are:

> Volition

> Opti-Jack

> MT-RJ (TE Connectivity version only)

Figure 2-7: Jack (Left) and Connector (Right)[1]

2.3.4 KEY

Keying is a design characteristic that was absent on the early designs, but is an requirement in current and future designs. The key prevents rotation of two ferrules relative to one another. Such prevention provides consistent power loss from connection to connection (Figure 2-8).

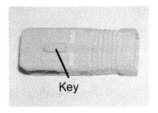

Key

Figure 2-8: Key On Connector

This variability, referred to as 'repeatability' is specified commonly as a maximum of 0.2 dB.[2]

This author has tested repeatability for both keyed and unkeyed connector styles. For keyed connectors, we find typical repeatability closer to 0.1 dB than to 0.2 dB. For unkeyed connectors, we find typical repeatability to vary from 0.5 dB to 1.0 dB.[3]

---

[1] Courtesy Panduit Corporation

[2] We obtained this value from a survey of connector data sheets.

[3] Data are from SMA 906 connectors tested from 1990-1994.

This improvement from keying is significant in two situations: initial network certification and maintenance or troubleshooting. During initial network certification, the lack of high repeatability of non-keyed connectors could result in the need to tune connectors to bring link power loss into compliance with requirement for the optoelectronic optical power budget available (OPBA, PFOI, 8.0). Subsequent disconnection and reconnection of any connector could result in loss of the preferred ferrule orientation and in link failure.

During maintenance or troubleshooting, the installer compares current loss measurements to original measurements. There is a possibility that improvement in the alignment of non-keyed connectors could mask a degradation of link components. With this possibility, interpretation of loss increase is more difficult with non-keyed connectors than with keyed connectors. With keyed connectors, any increase in loss greater than the range (PFOI, 14.6) Indicates degradation of the connectors or the link.

2.3.5 BACK SHELL

The back shell is the location to which a jacketed fiber is attached to the cable (Figure 2-9). The back shell design may, or may not, provide for axial load and lateral load isolations.

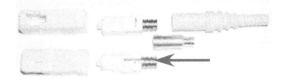

Figure 2-9: Connector Back Shell

2.3.5.1 AXIAL LOAD ISOLATION

Axial load isolation, otherwise known as 'pull proof' behavior, results from isolation of the ferrule from motion due to axial tension on the back shell. This isolation is achieved by inclusion of a spring between the ferrule and the back shell. Such isolation prevents an increase in power loss from tension applied to the cable.

2.3.5.2 LATERAL LOAD ISOLATION

'Wiggle-proof' behavior results from isolation of the ferrule from motion due to lateral pressure on the back shell. Such pressure will result in increased power loss if the ferrule tilts.

In summary, a pull-proof, wiggle-proof connector provides more reliable operation than one that is neither.

2.3.6 BOOT

The strain relief boot limits the bend radius of the cable or the tight buffer tube to increase the reliability of the connector (Figure 2-10). Forgetting to install the boot prior to connector installation is a common error. Evidence for this error is a split boot with tape!

Figure 2-10: Strain Relief Boot

2.3.7 CRIMP SLEEVE

The crimp ring, or crimp sleeve, grips the strength members of the cable (Figure 2-11). When making patch cords with 1.6-3 mm jacketed fibers, the installer uses this ring. When installing connectors on 900-μ tight tubes, the installer does not install this ring.

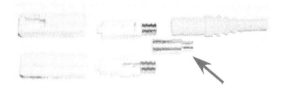

Figure 2-11: Crimp Sleeve

2.3.8 SIMPLEX AND DUPLEX

Most connector types are simplex. Some simplex connectors can be converted to duplex through the use of an external clip (Figure 2-12). If the mating mechanism is PO/PO, such conversion is possible.

Four connector types are simplex but can be converted to duplex: SC, LC, LX.5,

and MU. Five connector types are duplex by design: Volition, Fiber-Jack, FDDI, ESCON, and MT-RJ.

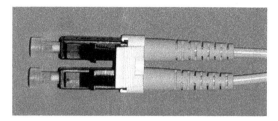

Figure 2-12: LO LC Duplex Connector

Duplex connectors for patch cords re recommended instead of two simplex patch cords. Use of duplex patch cords eliminates the possibility of crossing fibers. Crossed fibers are one of the most common errors in network installation and maintenance.

If the connector has a retaining ring, duplex conversion is not possible. Such connector types are: 905 and 906 SMA, biconic, mini- BNC, ST-™ compatible, FC, and D4.

2.3.9 FERRULE CAP

Connector caps or covers keep dust from the end of the fiber. Volition and LX.5 have caps that are integral to the connectors. All other connectors have caps that are separate and forgettable. Failure to install a cap on a connector results in dirt on the surface of the ferrule and core. Such dirt increases loss and can result in damage to the cores of mated connectors.

2.3.10 COLORS

TIA/EIA-568-C recommends the connector color-coding in Table 2-1. Connector bodies and/or boots can have the color. Barrels should have the same color as the connector to indicate the type of fiber behind the patch panel.

Color	Use
Green	APC singlemode
Blue	Singlemode
Aqua	50-μ LO
Black	50-μ
Beige	62.5-μ

Table 2-1: Connector Color Codes

2.4 PERFORMANCE

The installer has four connector performance concerns:

- Maximum insertion loss, dB/pair
- Typical insertion loss, dB/pair
- Repeatability, or range, in dB
- Reflectance, in - dB

2.4.1 DB PER PAIR

The unit of connector power loss is "dB/pair". Such loss is from one fiber to another. Thus, a physical pair is required. The term 'dB/connector' has no meaning.

By this definition, there is no power loss from a transmitter source to a fiber or from a fiber to a receiver detector. In an absolute sense, power loss occurs at both locations. However such loss is accounted for in the method of measurement of the optical power budget of the optoelectronics (PFOI, 8.0). Field measurements need not include this loss. Doing do results in double counting of this loss.

2.4.2 MAXIMUM LOSS

The maximum insertion loss, in dB/pair, is the maximum loss the installer experiences when he installs the connectors correctly. However, he should not expect to see this value, or a value close to this value, unless he makes errors during installation. Installers use this maximum value in certifying networks.

When the installer installs connectors correctly, the loss is due to *intrinsic* causes, such as core and cladding diameter variations, core offset, cladding non-circularity, NA mismatch, and offset of the fiber in the ferrule.

When the installer makes errors, he creates *extrinsic* causes of loss, such as damaged core, bad cleaves, dirt or contamination on the core, and air gaps due to excessive polishing.

All data communication connectors are rated at the maximum loss of:

- 0.75 dB/pair

This is a 'Magic Value' since it is always true. It applies to singlemode or multimode, keyed and contact connectors.

2.4.3 TYPICAL LOSS

The typical insertion loss, in dB/pair, is the value the installer experiences when he installs connectors correctly. Frequently, the installer experiences values lower than this value. The installer uses this value in certifying networks (PFOI, 19).

Singlemode and multimode, keyed and contact, connectors have three typical loss values ('Magic Values'):

- 0.15-0.20 dB/pair for connectors which require polishing and have 1.25 mm ferrules
- 0.30 dB/pair for connectors which require polishing and have 2.5 mm ferrules
- 0.40 dB/pair for connectors which require no polishing

2.4.4 REPEATABILITY

Repeatability is the maximum increase in loss, in dB, that a connector will exhibit between successive measurements. The repeatability is used to calculate the range when the installers have not done range testing as part of the installation testing (PFOI, 14.6). The range is the maximum increase in loss that occurs during successive measurement of the same link. If the installer has not performed range testing, he can calculate a range by doubling the repeatability.

2.4.5 REFLECTANCE

Reflectance is a measurement of the relative optical power reflected backwards from a connector. This reflected power can travel back to the light source and be reflected back into the fiber. If this reflected power reaches the receiver with a power level above its sensitivity, the receiver will convert this optical power to a digital 'one'. If the time interval in which the reflected power arrives was a 'zero' at the transmitter, the output signal will differ from the input signal. Thus, reflectance influences the accuracy of transmission of a fiber optic link. Minimizing reflectance maximizes signal accuracy. The objective of a reflectance requirement is to limit the reflected power at the receiver to a value less than the sensitivity of that receiver.

Reflectance occurs at a glass air interface. This interface, or any interface at which there is a change in speed of light, or index of refraction, can produce a reflection, called a "Fresnel reflection". This reflection occurs in fiber optic connectors because the end faces of mated connectors have surface roughness, which creates microscopic air gaps (Figure 2-13).

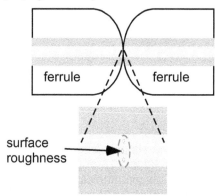

Figure 2-13: Air Gaps Create Reflectance

In connectors, this reflection is called reflectance. Reflectance is defined in Equation 2-1. Reflectance is stated in units of negative dB, with values from -20 dB to -65 dB. For reference, a non-contact connector exhibits reflectance between -14 dB and -18 dB.

Reflectance =

10 Log (reflected power/incident power)

Equation 2-1

Reflectance is described qualitatively by three terms:

> PC

> UPC

> APC

Commonly, PC (physical contact) refers to reflectance of less than –40 dB; UPC (ultra physical contact), to less than –50 dB; and APC (angled physical contact), to less than –55 dB.

Historically, reflectance has been a concern for singlemode connectors. However, at the time of this writing, reflectance is a concern in multimode networks transmitting at and above 1 gigabit per second. The gigabit Ethernet standard requires multimode connectors with reflectance less than -20 dB.

2.5 TYPES

We present connectors in three groups:

> Small form factor (SFF) connectors

> Commonly used connectors

> Legacy connectors

The SFF connectors are expected to become dominant in the future. The commanly used types are dominant at this time. The legacy connectors were used in the past but are not used in new installations. Unless otherwise stated, photographs in this section are approximately actual size.

2.5.1 SFF CONNECTORS

While the ST-compatible (2.5.2.1) and SC (2.5.2.2) connector types have had long, and successful market histories, both have the same, significant drawback: large size. This size results in increased optoelectronics cost. The large size of the connectors forces optoelectronic manufacturers to space transmit and receive electronics far apart. This large spacing results in half as many fiber ports in a fiber switch as in a UTP switch. This spacing increases switch cost.

In the early 1990s, the optoelectronic manufacturers requested a reduced size fiber connector that would enable them to reduce the per port cost of switches. Their goal was for cost parity of fiber switches with UTP switches.

Connector manufacturers answered this request with a series of small form factor (SFF) connectors. These SFF connectors obviously meet the need of reduced size (Figure 2-14).

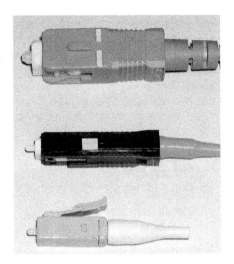

Figure 2-14: Size Comparison of Connectors: SC (top), MU (middle) And LC (bottom)

The connector manufacturers six answers were:

> LC

> LX.5

> MU

> Volition

> Opti-Jack

> MT-RJ

2.5.1.1 LC

The LC is a simplex SFF connector that can be converted to a duplex form with a clip. Lucent Technologies developed the LC (Figure 2-15 and Figure 2-12) as a connector to enable telephone companies to increase the density of installed connectors. Dense wavelength division multiplexing (DWDM) is the technology that created this need for increased density.

The LC is available from at least other six manufacturers. Many optoelectronics manufacturers provide products with the LC interface. These optoelectronics transmit at up to 10 Gbps.

The LC is available in radius and APC versions. The LC is a keyed, contact, low to moderate loss, pull proof and wiggle proof design. The LC has a 1.25 mm ferrule, which is half the diameter of the ferrules used in ST-compatible, FDDI and ESCON connectors. This small ferrule size can cause problems for installers, until they realize that half the diameter means one quarter of the usual polishing pressure.

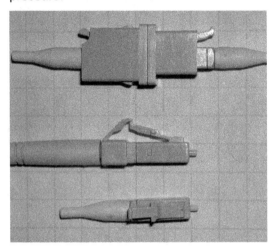

Figure 2-15: LC Simplex Connector

2.5.1.2 LX.5

The LX.5 is a simplex SFF connector that can be converted to a duplex form with a clip. ADC Telecommunications developed the LX.5 (Figure 2-16), similar to the LC, for use in telephone networks. ADC Telecommunications, now part of TE Connectivity, was the sole manufacturer. The LX.5, available in radius and APC versions, has a unique feature: a built in dust cover (Figure 2-16, bottom). This cover lifts as the connector is installed.

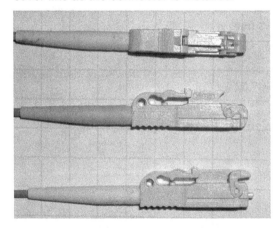

Figure 2-16: LX.5 Connector

The LX.5 is a keyed, contact, low loss, pull proof and wiggle proof design. LX.5 has a 1.25 mm ferrule.

As in the connector, the LX.5 adapter has a built in dust cover (Figure 2-17). The dust covers on the connector and in the adapter increase eye safety. With multiple wavelengths on singlemode fibers in DWDM networks, the total power level at the connector can be high enough to cause eye damage.

Figure 2-17: SC And LX.5 Barrels

2.5.1.3 MU

The MU is a simplex SFF connector with the appearance of an SC but with all dimensions reduced by half (Figure 2-18). Designed and developed in Japan, the MU is a keyed, contact, low loss, pull proof and wiggle proof design. It is unique in that it can be assembled to create a duplex, triplex or quadruple connector.

Figure 2-18: MU Connector

2.5.1.4 VOLITION™

The Volition™ duplex connector from 3M[4] is a plug and jack system (Figure 2-19, a jack open with darkened fibers in grooves, and Figure 2-20). The Volition incorporates a design feature that is both evolutionary and revolutionary: 'V'-grooves instead of ferrules.

V-grooves are evolutionary, in that they have been used for alignment of fibers in fusion splicers since the 1970's. V-grooves are revolutionary in that they have not been used in connectors prior to the Volition.

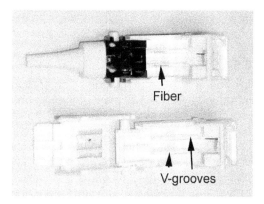

Figure 2-19: Volition™ Jack

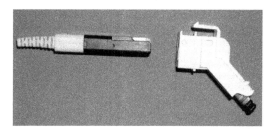

Figure 2-20: Volition™ Plug And Jack

3M used V-grooves for alignment of the fibers: in the jack, fibers rest in precision, molded, plastic V-grooves (Figure 2-19, top). In the plug, a custom fiber[5], floats in space. The plug fibers slide into the V-grooves of the jack, with contact of the mating fibers maintained through pressure created by bending of the fibers.

The Volition plugs and patch cords are factory assembled. The jacks are field installable onto a custom Volition cable. This custom cable is required, since the jack requires fibers with a primary coating of 250-µ. This diameter is smaller than the 900-µ tight tubes of commonly used premises cables. Because of this requirement, it is not possible to retrofit existing fiber optic networks with the Volition connectors.

In spite of this limitation on retrofitting, the Volition system proved popular with a limited number of facilities because it provided a low cost, fiber optic connection system. 3M achieved this low cost, approximately the same as that of a UTP

---

[4] The Volition™ connector is known as SG and VF-45.

[5] This fiber, known as GGP, for glass/glass/polymer, has a glass core, a glass cladding to a diameter of 110-µ and a hard plastic over-cladding to a diameter of 125-µ.

jack, by eliminating the single, largest cost in the connector- the ferrule.

2.5.1.5 FIBER JACK

In January 1997, Panduit Corporation introduced the first duplex SFF plug and jack connector system, currently known as a FJ® Opti-Crimp®, and Mini-Com® Duplex Jack Modules (Figure 2-7). This connector system, introduced a new package based on four existing and well proven technologies: 2.5 mm ferrules, quick cure adhesive installation, split alignment sleeves, and an RJ-45 form factor. Panduit's use of well-proven technologies eliminated the risk of using this 'new' product.

The 2.5 mm ferrules are the same as those in the ST-compatible, SC, FC, FDDI, and ESCON™ connectors. Quick-cure adhesives had been in use since 1992 and were well developed. The split alignment sleeves were commonly used in multimode adapters since 1986. The form was the same as that of the RJ-45, with which network installers and users were familiar.

By its appearance, the Opti-Jack™ is the most rugged of the SFF connectors and is well suited for fiber to the desk (FTTD) applications. The Opti-Jack™ is keyed, contact, moderate loss, pull proof and wiggle proof. It is a duplex connector with one ferrule per fiber. This structure allows separate alignment of each fiber for minimum power loss. Installation of the current product is by the cleave and crimp method (2.6.5).

2.5.1.6 MT-RJ

The MT-RJ was available in two versions, the plug and jack version from TE Connectivity (Figure 2-21) and the plug, adapter and plug version from Corning Cable Systems and others (Figure 2-22). Confirmed reports indicate that the Corning Cable System connector is no longer available.—The MT-RJ is keyed, contact, and pull proof, with two fibers in a single ferrule. Installation is by the 'cleave and crimp' method. At this time, MT-RJ optoelectronics are available at bit rates up to 1 Gbps. Photographs are 50 % actual size.

Although the MT-RJ plug has a size smaller than those of other SFF connectors, the MT-RJ density in patch panels is the same as that of other SFF connectors.

Figure 2-21: TYCO/AMP MT-RJ System

Figure 2-22: Corning Cable Systems MT-RJ System

2.5.1.7 E2000

The SI2000 is a European connector. It has an appearance similar to that of the LX.5. It has an integral dust cover (Figure 2-23). Typical loss is 0.12 dB/pair, with 97% of values at or below 0.25 dB/pair.

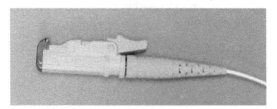

Figure 2-23: E2000 Connector

2.5.1.8   MTP/MPO

MTP/MPO connectors (Figure 2-24) have 12, 18, or 24 fibers in a single, molded polymer ferrule. The fibers can be ribbons. Some versions have up to 72 fibers in six rows of 12 fibers per row.

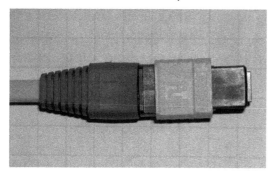

Figure 2-24: MTP® Connector[6]

This MTP is used with pre-terminated cable systems. Such systems are used in data centers. Such centers use pre-terminated cables, with modules on the ends. The modules break out the fibers to single fiber connectors, such as SC or LC. This cable and module system reduces installation time significantly. In addition, these connectors reduce the size of enclosures and patch panels.

The costs of pre-terminated systems vary, but are close to the those of field-installed systems. This author considers pre-terminated systems ideal for vertical riser backbones.

2.5.2   COMMON TYPES

The two most popular, or commonly used, connector types are the ST™-compatible[7] and the SC. Both have a 2.5-mm diameter ferrule.

2.5.2.1   ST-Compatible

Introduced in 1986, the ST™-compatible (Figure 2-25 and Figure 2-5) is a keyed, contact, moderate loss connector that is not pull proof or wiggle proof. Installation of the this connector is significantly easier and faster than that of predecessor types. The ST™-compatible connector requires more space on a patch panel than does

the SC. This space is required for rotation of the retaining ring. A consequence of this space requirement is increased patch panel and enclosure cost. In locations requiring more than 12 connectors, the ST™-compatible connectors can have a total installed cost higher than that of the SC connector.

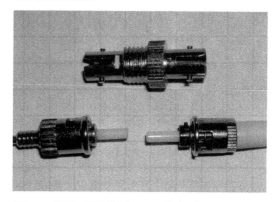

Figure 2-25: ST-™ Compatible Connector

2.5.2.2   SC

First available in the US in 1988, the SC (Figure 2-26 and Figure 2-27) was developed by Nippon Telephone and Telegraph (NTT) in Japan. The SC became commonly used after it became the connector type required for compliance with the Building Wiring Standard, TIA/EIA-568 (1990). Because of its exclusive recommendation in TIA/EIA-568 and TIA/EIA-568-A, the SC has dominated multimode data installations during 1995-2000.

The SC is keyed, contact, moderate loss, pull proof and wiggle proof. Because of these latter two characteristics, the SC is more reliable than the ST-compatible.

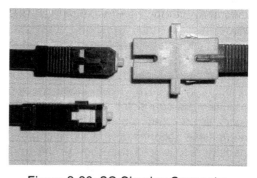

Figure 2-26: SC Simplex Connector

---

[6] Registered trademark of the US Conec Corp.

[7] ST is a trademark of Lucent Technologies.

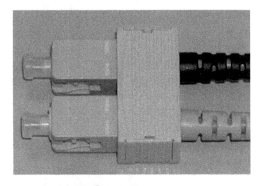

Figure 2-27: SC Duplex Connector

These pull and wiggle proof behaviors result in a price higher than that of the ST™-compatible. This behavior makes damage of the fiber difficult by controlling the force with which ferrules make contact. In contrast, the installer can apply excess contact force during the insertion of an ST™-compatible connector.

A second benefit of this behavior is immunity to damage from 'pull and snap'. An SC connector cannot be pulled and allowed to snap back into an adapter. An ST™-compatible connector can be Usually, such action results in damage to the fiber in an ST™-compatible connector.

We have observed this improved reliability: during fiber optic connector installation training, this author has experienced a damage frequency for ST™-compatible test leads that is 3-6 times that of SC leads. Thus, in spite of the increased price, the life cycle cost of the SC connector has been lower than that of the ST-compatible connector.

The SC differs from the ST-compatible in two additional aspects: insertion method and ability to be duplexed. The SC insertion method is 'push on/pull off'. Because of this method, the SC connectors can be more closely spaced in a patch panel than can the ST-compatible, which requires space for rotation. In practice, this reduced spacing results in a connector density of two to four times that possible with ST™-compatible connectors. This increased density results in a reduction in the total installed cost of SC connectors.

The SC can be duplexed, either through use of left and right outer housings with integral clips or an external clip (Figure 2-27). This capability results in increased network reliability because it is difficult to reverse the fibers in a duplexed SC connector. Thus, use of SC connectors results in three cost reductions:

> Life cycle cost

> Enclosure cost

> Maintenance and troubleshooting cost

2.5.3 LEGACY CONNECTORS

In this section, we present the legacy connectors. Legacy connectors are not used in new installations. These connectors are:

> SMA 905

> SMA 906

> Biconic

> Mini-BNC

> ESCON

> FDDI, MIC

> FC

2.5.3.1 SMA, BICONIC AND MINI-BNC

Early connector types, such as the SMA 905 (Figure 2-28), SMA 906 (Figure 2-29), biconic (Figure 2-30), and mini-BNC (Figure 2-31) had flat end faces and deliberate air gaps, which resulted in increased loss and high reflectance, −14 to -18 dB. These four types were non-keyed.

Both 905 and 906 were polished to a fixed ferrule length. The polishing tool controlled this length. The 905 and 906 differed in the use of a precision alignment sleeve (Figure 2-29). This sleeve was removable, replaceable, and forgettable! In addition, this author has observed wear after fewer than ten cycles. Wear produced small plastic shavings that blocked light.

The biconic controlled the ferrule length by polishing and measurement. This process was time consuming. The required tools were expensive.

Figure 2-28: 905 SMA Connector

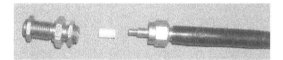

Figure 2-29: 906 SMA Connector

Figure 2-30: Biconic Connector

Figure 2-31: Mini BNC Connector

2.5.3.2 FC AND D4

Originally developed in Japan, the FC (Figure 2-32) and D4 (Figure 2-33) connectors have some of the characteristics of both the ST™-compatible and the SC connector types. Like the ST™-compatible, both have a rotational insertion method, which requires a relatively large spacing in a patch panel. Like the SC, both are keyed, contact, moderate loss, pull proof and wiggle proof. The FC/APC has a beveled end face and the same characteristics as the FC.

The D4 has a unique feature. During assembly, the key can be adjusted to achieve the lowest possible loss.

Figure 2-32: FC Connector

Figure 2-33: D4 Connector

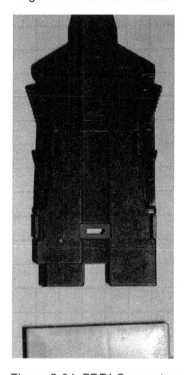

Figure 2-34: FDDI Connector

2.5.3.3 FDDI AND ESCON

All the previous legacy connectors are simplex. The last two, FDDI (Figure 2-34) and ESCON (Figure 2-35 and Figure 2-36) are duplex connectors. Both use the 2.5 mm ferrule. Both are large, keyed, contact, pull proof and wiggle proof. The FDDI connector is has a fixed shroud. In contrast, the ESCON connector has a movable shroud. The moveable shroud made cleaning ferrules is easy. In Figure 2-36, the shroud has been pushed back to reveal the ferrules. Both are considered obsolete. Photographs are 75% actual size.

Figure 2-35: ESCON Connector

Figure 2-36: ESCON Ferrules Exposed

2.6 INSTALLATION METHODS

Fiber optic connectors can be installed by many methods. These methods were developed to address different concerns.

Most of the development of methods has been to reduce one or more of the cost factors. In most cases, the effort to reduce the total installed cost resulted in increased connector cost. Five of these methods have become reasonably well received:

➢ Epoxy

➢ Hot melt adhesive

➢ Quick cure adhesive

➢ Crimp and polish

➢ No polish, no adhesive

In addition, we present two methods that may become more commonly used in the future:

➢ Fusion spliced pigtails

➢ Fuse on connectors

2.6.1 EPOXY

The epoxy, cure, and polish method was the first method used for installation of connectors. These epoxies can require different cures:

➢ A slow cure at room temperature

➢ A fast cure at room temperature

➢ A slow cure at elevated temperature

➢ A fast cure at elevated temperature

This method has four advantages:

➢ High resistance to degradation

➢ Loss stability over a wide temperature range

➢ High installation process yield

➢ Ability to be used with the lowest cost connectors

Epoxies are considered to have the high resistance to a wide range of conditions. As such, epoxies can provide the high reliability connectors. Other adhesive systems can have reduced resistance. As an example, connector epoxies can resist degradation to temperature of 105°C., while the original hot melt adhesive system was specified to 85°C.

Epoxies provide a good match of thermal expansion coefficients of epoxy, fiber and ceramic ferrules. This matching results in minimal relative movement of the fiber in the ferrule over a wide temperature range. Such limited movement results in stable power loss.

Use of epoxies results in high process yield. This high yield is a result of the bead on the ferrule (Figure 2-37). This bead supports the fiber during polishing. This support nearly eliminates damaged ends, which represent 95 % of connector losses during installation.

Figure 2-37: High Yield From Large Bead

Labor cost can be low for factory-installed connectors. The combination of low labor cost and low connector cost favors the use of epoxies as a factory installation method.

This method has three disadvantages.

> Inconvenient use

> Low installation rate

> Requirement for power for curing oven

The first two disadvantages are related as they translate to high installation hours per connector.

The use of epoxy requires mixing of two parts, transfer of the mixture into a syringe or automated injection mechanism, and clean up of excess epoxy. Because of these steps, the rate of installation of epoxy connectors is low, typically 6-9 per man-hour. This low rate results in a high total installed cost when field installation labor rates apply. Such rates exceed $30/hour. In comparison, factory installation labor rates can be $10/hour.

There is a second factor that increases the cost of epoxy connectors in field installation. That factor is labor utilization. Labor utilization is the fraction of the total time spent in installation by field installers. This utilization (also known as percent utilization) is much lower than that for factory installers: field utilizations can be 80% and below, while factory utilizations are above 90%.

Reduced labor utilization increases the cost impact of the low rate of installation

with epoxy. Because of the potentially high labor cost of this method, connector manufacturers developed other methods.

2.6.2   HOT MELT

Because of the inconvenience and time impact of epoxy installation, 3M developed the hot melt installation method. In this method, a hot melt adhesive is pre-installed into the connector. The installer preheats the connector to soften the adhesive so that he can install the fiber. The installer installs the fiber and/or cable into the connector and allows the connector to cool in a heat sink (a cooling stand). Once cooled, the installer removes the excess fiber and polishes the end in a two or three step polishing procedure. This installation method requires one or two 3M polishing films. These films are not easily clogged (i.e., worn out) by the hot melt adhesive.

The hot melt process eliminates the time factors of preparing and injecting the epoxy. In addition, this method eliminates the mess and inconvenience of epoxy. These eliminations allow an increased installation rate, 12-16 per hour. This increased rate can result in a reduced installation cost.

However, the hot melt connectors are more expensive than epoxy connectors. In addition, the hot melt method requires power for the heating oven and a proprietary oven, holders and polishing film. In spite of these factors, many installation situations can achieve a total installed cost with the hot melt method that is lower than that with the epoxy method.

A final advantage of the hot melt method is rework (7.10). With the ability to rework a damaged fiber end, the installer can achieve a 100 % yield. By itself, this ability can favor the hot melt method.

2.6.3   QUICK CURE ADHESIVE

Two manufacturers, Lucent Technologies and Automatic Tool and Connector, were the first to address the disadvantages of the proprietary parts of the hot melt method and the inconvenience of the epoxy method. They developed 'quick cure' adhesive methods.

The first advantage is elimination of the requirement of power. These methods, either two part adhesive (from Lucent Technologies and Automatic Tool and Connector) or one part (from TYCO), enabled an increase in the rate of installation. This increased rate, typically 15/hour, can result in reduced installation cost.

A second advantage of quick cure adhesives is their compatibility with two common ferrule materials: ceramic ferrules and some liquid crystal polymer (LCP) ferrules. This combination of low cost connectors with LCP ferrules and high installation rate can result in a low total installed cost. Installers can achieve this low installed cost if they achieve a high process yield. Yield is the ratio of the number of acceptable connectors to the total number of connectors installed.

There are three disadvantages of quick cure adhesives:

> Premature hardening of the adhesive

> Minimal support of the fiber during polishing

> Reduced reliability

The first two disadvantages contribute to a reduced process yield and increased cost. Premature hardening occurs when the fiber is coated with the hardener or accelerator and inserted into a connector loaded with adhesive. If the installer inserts the fiber slowly, the adhesive can cure and the fiber locks up before it is fully inserted. In this situation, bare fiber exists inside the connector. Such bare fiber can cause a reduction in the reliability of the connector.

For example, SC connectors allow the fiber inside the back shell to flex slightly during insertion into a patch panel or a receptacle. Repeated flexing of such bare fiber can result in fiber breakage. This author expects this failure to occur in any connector that is pull proof and wiggle proof, such as the FC, LC, and Opti-Jack.

This author has observed such failure in as few as 5 cycles.

The second disadvantage, minimal support of the fiber during polishing, occurs because the adhesive cannot produce a large bead on the tip of the ferrule. This inability is due to the nature of the adhesive: quick cure adhesives are edge-filling adhesives. Unlike the bead on epoxy or hot melt connectors, the small bead on the tip of the ferrule can allow the fiber to break below the surface of the ferrule during end finishing (Figure 3-18). Without extreme care during end finishing, the installation yield can be low.

For example, typical yield during training of novices with epoxy and hot melt connectors is 80-90%. Typical yield for these same novices with quick cure adhesives is 50-60%. However, typical yield of quick cure adhesive connectors for experienced professionals is 90-95 %.

The third, and final, disadvantage of the use of quick cure adhesives is reduced reliability. Some of the quick cure adhesives have exhibited degradation when exposed to a wide temperature range, a wide humidity range, rapidly changing temperature, or rapidly changing humidity. While there may be exceptions, quick cure adhesives are usually used for networks in office buildings.

2.6.4 CRIMP AND POLISH

Several manufacturers, AMP, 3M and Automatic Tool and Connector, have offered connectors that have a mechanical method of gripping the fiber. These products require polishing. These products are more susceptible to damage during polishing than are connectors installed with quick cure adhesives. This susceptibility to damage occurs because there is no bead supporting the fiber. Of the five methods presented herein, this method appears to have the lowest acceptance.

2.6.5 CLEAVE AND CRIMP

For the four installation methods presented, the major cause of reduced yield and increased cost is damaged or shattered fiber ends. This damage occurs during polishing. In addition, a significant amount of time is consumed by preparation, injection, and use of adhesives.

The 'cleave and crimp' installation method addresses both yield loss and high installation time by elimination of both adhesives and polishing. Such elimination offers the potential of increased installation rate and reduced installation cost. This increase in installation rate may be only a potential cost reduction, since the cost of these connectors is higher (2x-4x) than that of connectors installed by other methods.

'Cleave and crimp' is a generic term for no epoxy, no adhesive, no polish connector products with the trade names LightCrimp™ Plus (TYCO/AMP), Opti-Crimp® (Panduit Corporation), Unicam™ (Corning Cable Systems), and others.

All these products require the same steps:

> Cleave the fiber

> Insert the fiber

> Crimp the connector to the fiber

The connector contains a pre-installed fiber stub that has been polished by the manufacturer. In essence, this connector has a mechanical splice in its back shell.

The primary advantage of this installation method is reduced installation time, with the potential for reduced installation cost. Field installation rates can be 22-40/hour. Manufacturer literature includes claims that indicate rates higher than 40/hr. A second advantage is reduced training cost.

The realization of reduced installation cost depends on five factors. These five factors are:

> Total loaded hourly labor rate

> Labor time utilization

> Installation rate

> Installation yield

> Connector cost

These factors do not always combine to produce the lowest total installed cost. In spite of the complexity of the cost analysis, we can develop some rough guidelines.

> The higher the total loaded labor rate, the more likely the cleave-and-crimp method will result in the lowest total installed cost.

> As the number of connectors per location drops, the time utilization drops. As this utilization drops, the probability of lowest total installed cost increases for this method. Such situations include the desk locations in a fiber to the desk network (FTTD) but not a central equipment room in a vertical backbone or in an FTTD network.

This author's testing and use in training indicate that the cleave and crimp method yields and power losses are close to those of epoxy, hot melt and quick cure methods. However, the variability in loss (i.e., standard deviation) is higher than those of epoxy and hot melt methods.

2.6.6   PIGTAIL SPLICING

In the past, fusion splicer cost was much higher than at present. However, the availability of splicing machines in the $5000-$8000 range changes the total installed cost analysis of fusion spliced pigtails. If the total number of connectors is between 725 and 3000, the fusion-splicing machine will pay for itself in reduced installation cost. In other words, fusion spliced pigtails have the lowest total installed cost. After paying for itself, the splicing machine becomes an investment that continues to pay dividends. This analysis is available at: http://www.ptnowire.com /tpp-V3-l2.html.

2.6.7   FUSE ON CONNECTORS

The final installation method is that of 'fuse on' or 'splice on' connectors (SOC). These connectors have a pre-installed fiber with a fiber stub protruding from the back shell. The connector is fused to the fiber in the cable. Such products are available from Corning Cable Systems,

Alcoa-Fujikura, Sumitomo, Fitel, Clearfield, and Diamond.

Some of these products require use of a custom splicing machine. Others require use of a special holder with a standard machine. These latter machines have removable holders.

Such connectors are useful in situations of replacement or limited space. In replacement, the existing enclosure does not allow space for splice trays. In some situations, such as military ships, the space available does not allow for the space required by enclosures or splice trays. Such connectors have a cost higher than those of other installation methods.

All such products are one use. In other words, if the connector is high loss, either because of contamination of the cleave surface or because of a bad cleave, the connector is lost.

2.7　SUMMARY

In summary, the installer needs to know the following information about the connectors he is to install.

➢ Connector type

➢ Core diameter

➢ NA

➢ Maximum loss, dB/pair

➢ Average loss, dB/pair

➢ Repeatability, or range

➢ Reflectance

The installer needs to know the connector type, the core diameter and the NA to choose reference leads for testing. He needs to know the maximum and typical connector loss in order to calculate the acceptance value for certification of network (PFOI, 19).

He needs to know the range, or the repeatability, for troubleshooting. Finally, he needs to know the reflectance in order to certify the connectors.

2.8　REVIEW QUESTIONS

1.　True or false: contact connectors have higher loss than non-contact connectors.

2.　True or false: contact connectors have lower reflectance than non-contact connectors.

3.　True or false: contact connectors dominate today's market.

4.　True or false: non-contact connectors are common today.

5.　True or false: connector loss is a maximum of 0.75-dB/ connector.

6.　True or false: more reflectance is better than less reflectance.

7.　True or false: more reflectance provides lower BER than less reflectance.

8.　True or false: small form factor connectors are legacy connectors.

9.　True or false: SFF connectors drive network costs up.

10.　True or false: a D4 is a SFF connector.

11.　True or false: an MT-RJ is a simplex connector.

12.　What characteristic of the fiber contributes to the loss of non-contact connectors but not to that of connector connectors?

13.　List 8 characteristics of the connector structure.

14.　How many types of end faces are there?

15.　What are the types in Question?

16.　List 5 connector colors.

17.　What does each color indicate?

18.　What does SFF mean?

19.　The diameter of an SC ferrule is:

20. The diameter of an LC ferrule is:

21. Is an LC connector a simplex or duplex?

22. Is the MTP/MPO a simplex or duplex connector?

23. True or false: There are two ferrules in a Volition connector.

24. What type of connector is this?

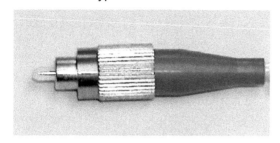

25. What type of connector is this?

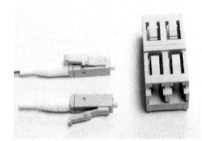

26. What type of connector is this?

27. What type of connector is this?

28. Why is the range of a connector important?

29. You observe a network technical start to install a green connector into a blue barrel. What should your response be?

30. True or false: when an installer installs connectors, he will often see the maximum loss value.

31. True or false: a normal experience is a measured value of 0.75 dB/connector.

32. Make a chart of all connector types presented in the text. On that chart, indicate the following: type; contact or not; latching mechanism, (PO/PO or ring); mating type, (plugs or plug and jack); end face type; simplex, duplex, or both; keyed or not; pull proof or not; wiggle proof or not.

33. What is the reason that the angle on an APC connector is 8°?

3 INSTALLATION PRINCIPLES

Chapter Objective: you learn the principles for installation of fiber optic connectors. These principles result in the procedures in Chapters 5-8.

3.1 INTRODUCTION

In this chapter, we present the principles that result in the procedures for connector installation by four methods:

- ➢ Epoxy and polish
- ➢ Quick cure adhesive
- ➢ Hot melt adhesive
- ➢ Cleave and crimp

These methods are commonly used and represent the majority of connectors installed in North America.

Taken as a group, these methods require combinations of the following six steps:

- ➢ End preparation of the cable
- ➢ Injection of adhesive
- ➢ Insertion of fiber into the connector
- ➢ Crimping
- ➢ Curing of adhesive
- ➢ End finishing

In this chapter, we differentiate between principles and methods. We indicate a principle with the symbol '▶'. We indicate the method of implementing the principle with the symbol '▶▶'.

3.2 CABLE END PREPARATION

Cable end preparation involves:

- ➢ Removal of the jacket(s)
- ➢ Removal of water blocking compounds
- ➢ Trimming of strength members
- ➢ Removal of buffer tube and primary coating (Figure 3-1, Figure 3-2, and Figure 3-3)

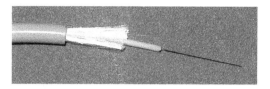

Figure 3-1: Single Fiber Prepared End

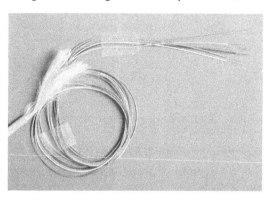

Figure 3-2: Eight Fiber Prepared Premises End

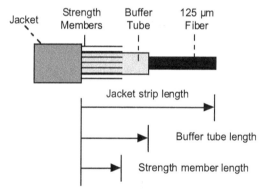

Figure 3-3: End Preparation Dimensions For Jacketed Fibers

3.2.1 DIMENSIONS

The cable end preparation dimensions depend on three factors:

- ➢ The nature of the installation
- ➢ The specific connector
- ➢ The specific enclosure into which the cable is to be installed

---

The end preparation can be either patch cord assembly (3.2.1.1) or enclosure installation (3.2.1.2).

3.2.1.1 PATCH CORD ASSEMBLY

For patch cord assembly, the connector determines the end preparation dimensions (Figure 3-3).

3.2.1.1.1 FIBER AND FERRULE

For connectors requiring polishing, two lengths, the jacket removal length and the bare fiber length, are long enough that:

> ➤ The bare fiber protrudes through the fiber hole in the ferrule

> ➤ The installer can remove excess fiber easily

The fiber must protrude through the ferrule so that the fiber can be polished flush with the ferrule to create a round, clear, featureless, flush core (4.4.1).

▶Fiber protrudes through ferrule

For epoxy connectors, the bare fiber length may be increased to allow easy removal of the excess fiber. However, if the bare fiber length is increased excessively, the installer may break the fiber while installing the connector into a curing oven.

For some connector styles, such as the SC and the LC, the jacket removal length is increased to allow a gap between the jacket end and the back shell of the connector. Such an increase enables a slight amount of flexing of the fiber under the boot. This flexing occurs when the connector is inserted into a patch panel. This flexing results in pull proof behavior (2.3.5.1 and 2.3.5.2).

Excessive jacket removal length can result in two problems, excessive bare fiber and short jacket. Excessive bare fiber length can result in breakage during insertion through the ferrule. In some cases, this broken fiber can become jammed in the fiber hole, resulting in loss of the connector.

When the connector has quick cure adhesive, excessive fiber length can result in premature adhesive curing and bare fiber inside the connector. Bare fiber

inside a connector results in reduced reliability.

The SC connector installed with epoxy or quick cure adhesive provides an example of such reduced reliability: if there is bare fiber inside the connector or if the buffer tube is short, so that the buffer tube is not immersed in the epoxy or adhesive, bare fiber remains inside the connector. The SC connector requires that the fiber in the back shell flex slightly whenever the connector is inserted into a patch panel. Repeated flexing of bare fiber results in breakage. During training programs, this author has observed such bare fiber breakage after 5-10 insertions.

If the jacket removal length is excessive, the boot will not control the bend radius properly. A cable that is not covered by the boot is a condition of reduced reliability (Figure 3-4 and Figure 3-5).

Figure 3-4: Proper Jacket Removal Length

Figure 3-5: Excessive Jacket Removal Length

3.2.1.1.2 STRENGTH MEMBER

The strength member must extend beyond the end of the jacket to be gripped by the crimp sleeve. For many connectors, the strength member length is 3/16" to 5/16".

Some connectors have no crimp sleeve. Instead the strength member folds back over the jacket, which fits into the back shell of the connector. The back shell of the connector is crimped or glued to both the strength member and the jacket.

Excessive strength member length can interfere with proper operation of the connector. For example, excessive strength member length will prevent sliding of the outer housing of an SC connector over the inner housing (Figure 2-3). Such sliding is necessary for both insertion and removal of the connector.

▶Strength member length must be proper

3.2.1.1.3 BARE FIBER

The buffer tube length must be long enough to butt against the inside of the ferrule, or to be completely immersed in an adhesive that hardens so the adhesive prevents motion of the bare fiber (Figure 3-6). Bare fiber in a connector is a condition of reduced reliability.

▶No bare fiber in connector

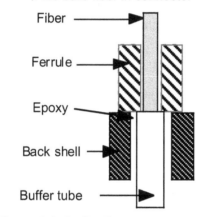

Figure 3-6: Buffer Tube Butts Against Ferrule

In order to comply with the principles in 3.2.1.1.1-3.2.1.1.3, the installer takes two steps:

▶▶Determines dimensions

▶▶Uses template

He uses a template for each connector type to control the stripping dimensions. This template has the form of Figure 3-3.

3.2.1.1.4 CRIMPING

This section applies to connectors that require adhesive. Connectors that require no adhesive and fiber cleaving require a crimper unique to the connector. Crimping of such a connector grips fiber and buffer tube.

The installer performs crimping to:

➢ Attach the cable strength members to the connector back shell

➢ Grip the jacket

3.2.1.1.5 CRIMPER NEST

The tolerance of the crimp sleeve is tight. An oversized crimp nest will not grip the strength members, allowing the connector to be pulled from the cable easily. An undersized crimp nest destroys the crimp sleeve. For example, a 0.141" crimp nest may result in a loose crimp sleeve, while a 0.137" nest results in a tight sleeve. In crimping, close is not good enough!

To avoid such problems, the installer uses the exact crimp nest size(s) recommended by the connector manufacturer (Figure 3-7).

▶▶Use proper nest size

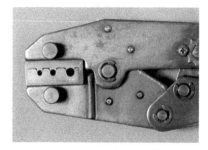

Figure 3-7: Crimp Nest In A Crimper

The large diameter of the crimp sleeve attaches the strength members to the connector. Once the crimp sleeve is attached to the connector, the connector cannot fall off the cable. The small nest attaches the jacket to the connector, so that the jacket cannot shrink away from the connector.

▶▶Large crimp first

3.2.1.2 ENCLOSURE INSTALLATION

For installation of connectors onto a cable that is installed in an enclosure, both the enclosure and the connector determine the end preparation dimensions. These dimensions are determined by four principles:

▶Sufficient buffer tube length

▶Strength member length proper

▶No bare fiber

►Fiber protrudes through ferrule

First principle: for all connectors, the jacket removal length is long enough to enable the cable to be attached to the enclosure at a specific location (Figure 3-8 and Figure 3-9), to allow the buffer tube to extend outside the front of an enclosure for connector installation, and to allow the end of the buffer tube to reach the patch panel. The buffer tube length will be coiled inside of the enclosure (Figure 3-10).

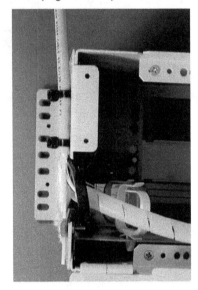

Figure 3-8: Indoor Cable Attachment Location

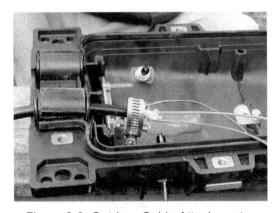

Figure 3-9: Outdoor Cable Attachment Location

This length must not be excessive, as excessive length can result in damage to the buffer tube when the volume of buffer tubes exceeds the capacity of the enclosure. A practical tolerance on buffer tube length is ±1-2".

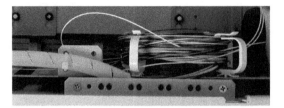

Figure 3-10: Buffer Tube Coiled Inside Indoor Enclosure

Second principle: the enclosure determines the strength member length. The enclosure is designed so that the strength members can be attached at a specific location (Figure 3-8 and Figure 3-9). In general, slightly excessive strength member length is a cosmetic or workmanship concern, but not a performance or reliability concern.

► Strength member length must be sufficient

Third principle: there is no bare fiber inside the connector, is almost automatic. Without a jacket, there is nothing to prevent the fiber from entering the connector until the buffer tube butts against the inside end of the ferrule.

The fourth principle, fiber protrudes through ferrule, controls the bare fiber length. The fiber will protrude beyond the tip of the ferrule, as long as the installer

►►Determines dimensions

►►Uses template

3.2.2 FIBER PREPARATION

In order to install connectors, the installer avoids damaging and contaminating the cladding.

3.2.2.1 CLADDING DAMAGE

Because the installer removes the primary coating, the cladding can become scratched or otherwise damaged. Such damage can result in fiber breakage. The installer

►Avoids damage

Avoiding damage requires cleaning with lens grade tissues, which are not abrasive.

►►Clean fiber with lens grade tissues

If the cladding is exposed to air for a significant length of time, moisture in the air can reduce the fiber strength. By itself, water will not reduce fiber strength. Moisture in the air is not pure water. As such, it can attack the fiber and reduce its strength. Thus,

> ▶▶Strip, clean and insert fiber without delay

3.2.2.2   CONTAMINATION

Low power loss requires a small clearance between the cladding and the inside of the fiber hole. This small clearance requires prevention of particulate contamination of the cladding surface.

> ▶Avoid contamination

A second form on contamination is chemical contamination. To avoid this, the installer

> ▶▶Uses 98 % isopropyl alcohol

70 % isopropyl alcohol contains water and oil. Both will contaminate the cladding with a potential to prevent adhesion of an epoxy or adhesive.

The installer cleans the fiber with lens grade, lint free tissues to remove contamination. After cleaning the fiber, the installer does not place it down or against any surface. To avoid contamination, the installer

> ▶▶Inserts the fiber in the connector immediately after cleaning

3.3   ADHESIVES

We use the term 'adhesive' to mean any material that is used to glue, or fix in place, the fiber inside the connector. Fiber optic connectors use three adhesive types:

- ➢ Epoxy
- ➢ Quick cure adhesive
- ➢ Hot melt adhesive

3.3.1   EPOXY

We define the term 'epoxy' to mean a one or two part chemical system that requires

a minimum cure time of two minutes with heat or 6 hours without heat.

3.3.1.1   THREE FUNCTIONS
Epoxy provides three functions:

- ➢ A strong bond between the fiber and the ferrule
- ➢ A strong bond between the buffer tube, or primary coating, and the connector back shell
- ➢ Support of the fiber protruding beyond the end of the ferrule during polishing

3.3.1.2   TYPES
We organize fiber optic epoxies into two main groups, each of which has three subgroups. The two main groups are:

- ➢ Heat cured epoxies
- ➢ Room temperature cure epoxies

The three subgroups are based on curing time:

- ➢ Fast cure (1-5 minutes)
- ➢ Medium cure (5-60 minutes)
- ➢ Slow cure (> 60 minutes)

Heat reduces cure time, but increases the chance of thermal cracking of the fiber. As cure temperature increases, the likelihood of thermal cracking increases. At the time of this writing, thermal cracking is rare.

3.3.1.3   EPOXY STRENGTH
The curing time and temperature must be sufficient to achieve adequate strength between the fiber and the ferrule. With insufficient strength, tension on the fiber can result in the fiber withdrawing into the ferrule. Alternatively, pressure on the fiber can result in the fiber protruding from the ferrule. Both types of motion, called 'pistoning', are undesirable.

> ▶Sufficient strength required

The combination of curing temperature, time, and epoxy results in sufficient strength.

> ▶▶Use temperature and time appropriate for the epoxy

Heat curing can be used, as long as the combination of temperature, fiber core diameter, ferrule material, and epoxy does not result in cracking.

Cracking can result when a heat cured connector cools to room temperature. The different thermal expansion rates of fiber, epoxy and ferrule can result in excessive compression of the fiber. This compression causes cracking. Cracks in fibers can divert the light from its proper path, causing high loss. At the time of this writing, thermal cracking is uncommon. The installer avoids cracking either by

▶▶Using proper combination

▶▶Using no heat

Whenever practical, the installer uses an epoxy that cures at room temperature. Such an epoxy will not exhibit thermal cracking. In addition, there will be no thermostat to malfunction. Such a malfunction can result in cracking.

Of course, there are situations in which use of heat is desirable. Small field installations and cable assembly operations use heat to reduce cure time.

3.3.1.4   INJECTION

The installer injects epoxy into the back shell of the connector with a syringe. The amount he injects is critical to proper operation of the connector.

3.3.1.4.1   EPOXY INTERFERENCE

The epoxy is injected so that the epoxy does not interfere with proper operation of the connector.

▶No interference

Excessive epoxy can create connector malfunction in two ways:

➢ Expansion

➢ Displacement

When heated, excessive epoxy in the back shell will expand. Such expansion can cause the epoxy to flow out of the back shell onto other areas of the connector. Such epoxy can cause the connector to fail to function properly.

When the fiber and buffer tube are inserted into the back shell, excess epoxy

is displaced from the back shell. In the case of the SC and LC connectors, such displaced epoxy can flow between the central tube and the back shell. In this case, the connector will lose its pull proof and wiggle proof performance.

For example, the SC and LC connectorscontains a tube inside the back shell (Figure 3-11 and Figure 3-12). This tube is mechanically independent of the back shell. Should epoxy flow between the tube and the back shell, the connector will lose its pull proof and wiggle proof performance (2.3.5.1 and 2.3.5.2). In addition, excessive, expanding epoxy can wick under a jacket via the strength members, making the cable rigid and sensitive to handling.

Figure 3-11: SC Inner Tube

Figure 3-12: LC Inner Tube

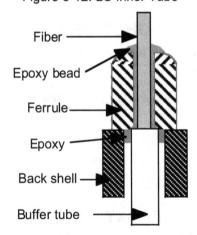

Figure 3-13: Sufficient Epoxy In Back Shell

To avoid excessive epoxy, the installer injects enough epoxy to fill the fiber hole in the ferrule and one additional drop in the bottom of the back shell (Figure 3-13). The additional drop bonds the buffer tube to the connector. To avoid interference, the installer

▶▶Minimizes epoxy

3.3.1.4.2   BEAD SIZE

Epoxy connectors have a bead on the ferrule tip (Figure 3-13). During polishing, this bead supports the fiber. With this support, fiber shattering and mechanical cracks from polishing can be eliminated.

The bead size determines polishing time and process yield through avoidance of shattering. The benefit of increased bead size is reduced incidence of shattering and cracking. However, as bead size increases, polishing time and cost increase. Obviously, there is a trade off.

This trade off depends on the installer's experience: the more experience the installer has, the smaller the bead can be.

A practical strategy is a bead of approximately 0.020-0.030" high (Figure 3-14) for novice installers and a 0.015" bead (Figure 3-15) for experienced installers. As a visual reference, installers can use a paper clip. Paper clips have a diameter of approximately 0.030".

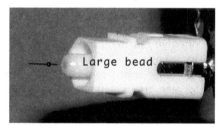

Figure 3-14: Large Bead

Figure 3-15: Small Bead

▶Control the bead size

There are three methods of controlling bead size. The first is to inject epoxy until it appears through the fiber hole in the ferrule. Immediately, the installer removes the syringe from the connector. When the installer inserts the fiber through the ferrule, the fiber will force additional epoxy onto the ferrule tip. This method results in a functional bead size.

The second method requires the installer to wipe the epoxy from the tip of the ferrule before inserting the fiber. The size of the bead will be controlled by the epoxy forced through the fiber hole by the fiber. As long as the bare fiber length is consistent, this method results in a consistent bead size.

The third method requires insertion and retraction of the fiber until it is flush with the ferrule. The installer wipes all epoxy from the tip of the ferrule and reinserts the fiber. This method results in a small bead, which is appropriate for hand polishing by experienced installers.

Inexperienced installers can add epoxy to the tip of the ferrule. This additional epoxy enables polishing without shattering or cracking the fiber.

3.3.1.5   CURING

Epoxy curing has three variables: expiration date, time, and temperature. Epoxies may not cure properly after their expiration date. In addition, some epoxies will not cure if they have been exposed to freezing temperatures.

Epoxies will cure properly as long as the curing time exceeds a minimum time. Additional time will not cause degradation of the bond.

Temperature is a different matter. Low cure temperature will not result in full bond strength between the fiber and the ferrule. Without such a bond, the fiber can move into and out of the ferrule ('pistoning'). Such movement can result in link failure due to insufficient power at the receiver. Excessive curing temperature can result in degradation of the bond strength and in fiber cracks.

The installer controls the temperature at that recommended by the epoxy manufacturer. To ensure such control, he monitors the oven temperature with an accurate thermometer. To ensure proper curing, the installer:

> ▶▶Controls oven temperature

> ▶▶Monitors oven temperature

> ▶▶Cures for at least the minimum time

To ensure the best possible reliability, the installer chooses an oven with a thermostat that fails in the 'full off' mode. An oven without such a thermostat will overheat and can crack fibers.

3.3.2 QUICK CURE ADHESIVES

We define 'quick cure adhesive' to mean a one or two part chemical system that cures without heat in a relatively short time, typically less than two minutes.

These adhesives have expiration dates. After the expiration date, the adhesive cures slowly. Eventually, the adhesive fails to cure. Therefore, prior to use, the installer:

> ▶▶ Checks the expiration date

3.3.2.1 TYPES

Quick cure adhesives can be a two part liquid product or a one-part gel product. In general, both are edge-filling adhesives. Edge-filling adhesives harden when the adhesive fills a narrow space, such as that between the outside of the fiber the inside of the fiber hole in the ferrule.

3.3.2.2 INJECTION

Excess quick cure adhesive inside the connector creates the same problems as does excess epoxy. In addition, two part, quick cure adhesives do not cure inside the back shell. Thus, excess adhesive inside the connector has no benefit.

By itself, a two part, quick cure adhesive can cure slowly. As a result of this characteristic, the installer need not rush the insertion of the fiber after adhesive injection out of concern for premature curing. After being injected, the adhesive

may not cure for five minutes. Therefore, the installer

> ▶▶Need not rush fiber insertion after adhesive injection

3.3.2.3 PRIMER APPLICATION

The second part of a two-part, quick cure adhesive system, is called a primer, accelerator, or hardener. This part is applied to the fiber prior to its insertion into the connector. This application can be by spraying, dipping or brushing. The best method of application is

> ▶▶Brushing of primer

We recommend against spraying, as spraying can create an acetone mist, a carcinogen.

Dipping would seem to provide the most consistent and complete coverage of the fiber. However, dipping is not convenient when the primer bottle is partially empty. Brushing has proven convenient and without any problems.

3.3.2.4 FIBER INSERTION

After brushing the primer onto the fiber, this author has observed a problem with insertion of the fiber. If the time between application of the primer and insertion of the fiber is excessive, the fiber may not fit into the fiber hole. After brushing the primer onto the fiber, the installer:

> ▶▶Inserts the fiber without delay

The adhesive starts curing as soon as the fiber is inserted. Quick-cure adhesives cure in less than two minutes, and, occasionally, in as little as 30 seconds. Because of this short cure time, the installer inserts the fiber into the connector quickly. To do otherwise would allow the adhesive to cure prior to full insertion of the fiber. Premature curing results in bare fiber inside the connector, a condition of low reliability. To avoid premature curing of the adhesive, the installer

> ▶▶Primes and inserts the fiber without delay

3.3.2.5   TIP BEAD

Insertion of the fiber forces the quick cure adhesive from the fiber hole onto the ferrule tip. As the adhesive cures only when in thin sections, application of the primer to the fiber or to the tip of the ferrule results in a very small bead (Figure 3-16).

Repeated application of adhesive and primer to the tip of the ferrule does increase the bead size. Increased bead size increases the resistance to fiber breaking below the surface of the ferrule during polishing (Figure 3-17). However, the increase in bead size is not significant. In addition, application of additional adhesive to the ferrule tip increases connector installation time, thus reducing the advantage of this method.

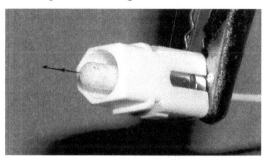

Figure 3-16: Bead Without Application Of Additional Adhesive

Curing begins as soon as the fiber is inserted into the connector. The older the adhesive is, the longer the cure time will be. Since 'old' adhesive cures slowly and provides increased time for fiber insertion, the installer in training can

▶▶Use old adhesive

Figure 3-17: Bead Size With Three Adhesive Applications

This author's experience is that quick cure adhesives can be used to several years

after the expiration date. Eventually, such adhesive will fail to cure.

To verify full curing, the installer pulls on the cable. If the fiber protruding beyond the tip of the ferrule does not move into the ferrule, the adhesive has cured fully.

3.3.3   HOT MELT ADHESIVE

We define the term 'hot melt adhesive' to mean a one part adhesive with two characteristics: it is preloaded in the connector and requires connector pre-heating prior to insertion of the fiber. 3M pioneered and, at the time of this writing, is the sole source of this system.

3.3.3.1   ABILITY TO REHEAT

Because the adhesive can be reheated, it is possible to salvage and repair damaged connectors. This characteristic results in reduced installation and maintenance costs. Such repair is possible under two independent conditions:

➢ Reheating must be done prior to crimping of the crimp sleeve

➢ There is extra fiber inside the connector

To provide extra fiber inside the connector, the installer inserts the fiber fully into the connector. He withdraws the fiber by approximately 1/16". After the installer re-heats the connector, he can push this extra fiber through the ferrule for a second polishing to repair a damaged core.

▶▶Leave extra fiber

If a fiber is so badly shattered that it cannot be re-polished, installer can reheat the connector, remove the fiber, prepare a new cable end, and reinsert the new end. This method can result in 100% yield and reduced maintenance cost.

This method works well with connectors installed on premises cable. This method can work once or twice with jacketed cable and the ST™-compatible connector. This method will not work when installed on a jacketed cable with the SC connector.

3.3.3.2   POLISH TIME

The viscosity of the hot melt adhesive is so high that it can stick to the fiber as the installer inserts the fiber. The experienced installer can reduce polishing time by increasing the bare fiber length by 1/2". By increasing the fiber length, the installer reduces the bead size and polishing time to 20-30 seconds for multimode connectors. However, if the fiber length is excessive, there will be so little adhesive on the tip of the ferrule that any fiber breakage during scribing or polishing will require connector rework or replacement.

To reduce polishing time, the installer can

▶▶Increase bare fiber length

3.3.4   FIBER INSERTION

Regardless of the method of installation, the installer inserts a fiber into a connector. Low loss requires a small clearance between the cladding and the fiber hole in the ferrule. In addition, the clearance in singlemode connectors is smaller than that in multimode connectors. The installer will exercise more care inserting fibers into singlemode connectors than into multimode connectors.

3.3.4.1   FIBER BREAKAGE

The installer's objective is insertion of the fiber without breakage. If the fiber breaks, the installer may loose the connector. If the fiber does not bend, it cannot break. Therefore, during fiber installation, the installer

▶Does not bend the fiber

To avoid bending during fiber insertion, the installer

▶▶Rotates the connector back and forth

Rotation of the connector allows the fiber to slip past lips or steps inside the connector.

This rotation has a second benefit: centering of the fiber in the fiber hole. The viscosity of the adhesive can cause the fiber to be better centered in the fiber hole than if the fiber in inserted without rotation.

Difficulty during fiber insertion may be due to dirt in the fiber hole. To clean dirt from a fiber hole, the installer can

▶▶Flush the connector with a syringe filled with isopropyl alcohol

3.4   END FINISHING

End finishing is the process of removing excess fiber from the end of the ferrule and creating an optical grade surface on the fiber end. When properly performed, this process results in a core that is round, clear, featureless and flush with the surface of the ferrule (4.4.1). This process requires three steps:

➢ Excess fiber removal

➢ Air polishing

➢ Pad polishing

3.4.1   FIBER REMOVAL

The objective of fiber removal is removal of excess fiber without causing it to break below the surface of the ferrule (Figure 3-18 and Figure 3-19). If the fiber breaks below this surface, the installer may not be able to polish the fiber so that the core is defect free and low loss. The installer removes the fiber by scribing it.

▶Fiber removed above ferrule

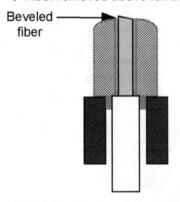

Beveled fiber

Figure 3-18: Fiber Broken Below Ferrule

Scribing is the placement of a single, small scratch on the fiber at the tip of the bead of epoxy or adhesive (Figure 3-20). Scribing is not sawing the fiber or breaking the fiber.

▶Scribing is scratching, not sawing

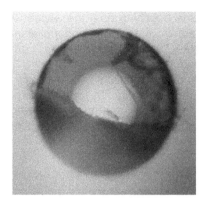

Figure 3-19: Broken Fiber Appearance

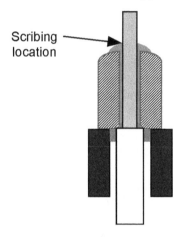

Scribing location

Figure 3-20: Location Of Scribe

The installer uses a scriber with a wedge shape (Figure 3-21), not one with a point. The installer can align the edge of a wedge scriber to the fiber more easily than the point of a 'pencil' type scriber.

▶▶Use wedge scriber

During scribing, the installer rests his hands together. By resting his hands together, the installer will not accidentally break the fiber with the scriber. If he hits the fiber, he can break the fiber below the surface of the ferrule (Figure 3-18).

Figure 3-21: Wedge Scribers

▶▶Scribe once

The installer scratches the fiber once, and only once, with a light pressure. With a single scratch, the fiber breaks on a single plane. If the fiber breaks on multiple planes, the fiber may break below the surface of the ferrule.

After scribing the fiber, the installer pulls the fiber away from the tip of the ferrule. To do so without breaking the fiber below the ferrule, the installer must not bend the fiber to the side. To avoid bending, the installer slides his fingers up the ferrule, onto the fiber. He pulls the fiber along its axis. In short, to avoid fiber breakage below the surface of the ferrule, the installer will

▶▶Pull the fiber

3.4.2 AIR POLISHING

The goal of air polishing is creation of a fiber end that is flush with the bead of epoxy or adhesive (Figure 3-22) and has no sharp edges (Figure 3-23). If the installer achieves both of these characteristics, the fiber will not snag on the polishing film during pad polishing. If he does not snag the fiber, he will not break it below the end of the ferrule. Air polishing requires rubbing of a relatively coarse polishing film against the fiber. No pad is used. The principle of air polishing is:

▶Flush fiber, dull edge

The rule is:

▶▶Always air polish before pad polishing

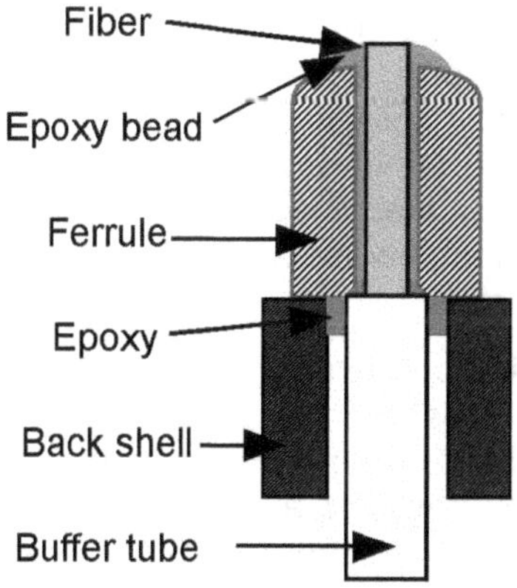

Fiber
Epoxy bead
Ferrule
Epoxy
Back shell
Buffer tube

Figure 3-22: Desired Air Polish Condition

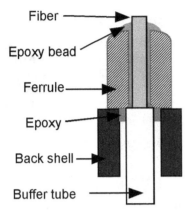

Figure 3-23: Undesired Condition

3.4.3 PAD POLISHING

3.4.3.1 OBJECTIVE

Pad polishing has two objectives:

> Creation of a lens grade, or optical grade, surface on the fiber end

> An end that is flush with the ferrule

Polishing is performed on rubber pads to allow the fiber to conform to the radius of curvature of the ferrule (2.3.1.2), resulting in low loss and low reflectance.

Glass and hard plastic polishing plates produce a flat fiber end, high loss and high reflectance. Such plates are no longer used except on legacy non-contact connectors, such as biconic and SMA connectors.

3.4.3.2 TYPES

There are two types of pad polishing: hand polishing and machine polishing. Hand polishing results in the desired low loss. Cable assembly facilities use machine polishing in place of hand polishing, as it results in reduced cost. In addition, machine polishing results in consistent low reflectance on singlemode connectors.

Machine polishing of up to 24 connectors to low loss and low reflectance can take as little as three minutes. In comparison, field polishing of a single connector can take thirty seconds (multimode) to three minutes (singlemode).

Machine polishing has one aspect that is different from that of hand polishing. All machine polishing requires grinding through the ferrule to produce a round cladding. Hand polishing may not produce a round cladding. In order to remove ceramic ferrule material, all machine-polishing films, except the final film, are diamond films.

3.4.3.3 FILMS

Both hand and machine polishing are done with relatively fine abrasive film(s) placed on a resilient rubber pad.

Fine films have grit sizes from 5-μ to 0.05-μ. This author's experience is that:

> Low loss and good appearance on multimode connectors require that the final film be at least as fine as 1μ

> Low loss, good appearance, and low reflectance on singlemode connectors require a final film at least as fine as 0.5μ

Some singlemode polishing procedures require a polish solution, which contains fine, suspended abrasives.

All polishing films are harder than the fiber. Diamond films are harder than the hardest ferrule material, which is ceramic. Alumina (aluminum oxide) films are harder than LCP, or composite, and stainless steel ferrules, but softer than ceramic ferrules.

Diamond film polishing removes fiber and ferrule. The fiber stays nearly flush with the ferrule (Figure 3-24).

Polishing ceramic ferrules with alumina films removes fiber but not ferrule. Excessive polishing with alumina film results in undercutting, which creates an air gap, high loss, and high reflectance (Figure 3-25).

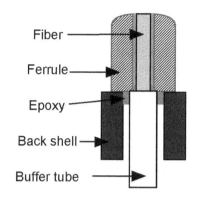

Figure 3-24: Fiber Flush With Ferrule

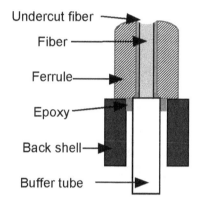

Figure 3-25: Undercut Fiber

3.4.3.4   POLISHING TOOL

The installer aligns the connector to the pad with a polishing tool, also known as a puck and fixture (Figure 3-26).

Figure 3-26: Polishing Fixtures

3.4.3.5   POLISHING LIQUID

The installer can perform wet or dry polishing. Cable assembly polishing is usually wet. For wet polishing, the installer uses distilled or reverse osmosis (RO) water. These types of water are free from particles that can cause scratches

on the core. Such scratches divert the light from its proper path and can result in increased loss and reflectance. In addition, wet polishing increases film life, as the liquid flushes polishing materials from the films.

▶▶Use distilled or RO water for wet polishing

In environments with dirty air, the installer uses dry polishing. Many field environments have significant amounts of particles in the air. In such environments, wet polishing is not the best choice, as the liquid can attract particles from the air. These particles can cause scratches on the core. This author's recommendation is to

▶▶ Use no liquid for field polishing

3.4.3.6   EQUIPMENT PREPARATION

One objective of polishing is the removal and avoidance of scratches. Scratches can be from contamination or from previous polishing steps. Scratches on the core can divert and block light from its normal path, resulting in increased connector loss.

▶Avoid and remove scratches on the core

Prior to use, the installer cleans the connector, the polishing pad, the polishing films, and tool to avoid scratching the core.

▶▶Prior to use, clean equipment

In addition, the installer cleans the connector and fixture before he changes to a film with grit finer than that of the previous film.

▶▶Clean connector and tool when changing to finer film

The installer can clean with lens grade tissues and one of the following: lens grade gas, distilled water or 98% isopropyl alcohol. Gas is convenient, but relatively expensive. Isopropyl alcohol is less convenient but relatively inexpensive.

3.4.3.7   POLISHING MOTION

The installer polishes the connector with a 'Figure 8' motion. This motion avoids creation of a bevel on the end of the fiber (Figure 3-27). Such a bevel will create an air gap between mated connectors. An air gap results in high loss and high reflectance.

▶▶Use 'Figure 8' polish

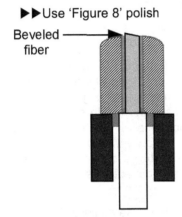

Figure 3-27: Bevel On End Of Fiber

3.4.3.8   POLISHING PRESSURE

The installer performs polishing without breaking the fiber below the surface of the ferrule. Should the bead of epoxy or adhesive be sheared from the tip, the fiber will not conform to the radius.

▶▶Use light polishing pressure on bead

At the start of polishing, the fiber may protrude above the bead on the tip of the ferrule (Figure 3-23). This condition will be evident by a 'scratchiness' of the connector on the film. As long as scratchiness continues, the installer will use a very light polishing pressure, a half-inch high figure 8 pattern, and a slow polishing motion.

▶▶Scratchiness requires slow, light pressure polishing

The installer continues with a light polishing pressure and slow polishing motion until the fiber is flush with the bead. This condition is evident by a lack of scratchiness. If the installer does not detect the end of scratchiness, the fiber is flush when the bead is gone. The bead is gone when the tip of the ferrule is mirror like.

Hot melt connectors require less pressure than do epoxy connectors. Hot melt adhesive is gummy or rubbery. If the

installer uses excessive pressure during polishing, the adhesive compresses but the fiber does not. In this case, the fiber can protrude beyond the bead of adhesive (Figure 3-23), snag on the polishing film, and break below the surface of the ferrule (Figure 3-18 and Figure 3-19).

Hot melt and quick cure adhesives can create a very small bead on the tip of the ferrule. A small bead provides less support of the fiber during polishing than does a bead of epoxy. Because of this difference in support, the installer performs the first pad polishing with less pressure than he would use on an epoxy connector. This reduced pressure is required until the fiber is flush with the ferrule. Once the fiber is flush, the installer cannot shatter the fiber below the ferrule.

▶▶Use reduced pressure on hot melt and quick cure connectors

Connectors with the 1.25 mm ferrule (LC, LX.5 and MU) have a cross section area that is 25 % of that of 2.5 mm ferrules. In order to avoid excessive pressure and broken fibers, the installer reduces his polishing pressure to 25% of the pressure he would use when polishing the 2.5 mm ferrules.

▶▶Use reduced pressure on 1.25 mm ferrules

3.4.3.8.1   POLISHING TIME

When polishing ceramic ferrules with films other than diamond, the installer polishes for a minimum time. Over polishing with alumina films can result in undercutting, high loss and high reflectance (Figure 3-25).

▶▶Minimize polish time

3.4.3.9   REPOLISHING

To remove minor damage from the core of a fiber in a ceramic ferrule, the installer uses diamond-polishing films. These films remove damage by removing both fiber and ferrule.

▶▶Salvage ceramic ferrules with diamond films

To remove minor damage from the core of a fiber in an LCP or stainless steel ferrule, the installer can use either diamond or alumina polishing film. Because alumina films are much less expensive than diamond films, they are preferred for salvaging connectors with these two ferrule materials.

▶▶Salvage other ferrules with alumina films

Note that it is possible to salvage ceramic ferrules with alumina films. However, the number of polishing strokes is limited by avoidance of undercutting. In other words, polishing with alumina may work. However, if the number of strokes is excessive, undercutting, high loss and high reflectance result.

3.5 CLEAVE AND CRIMP INSTALLATION

The 'cleave and crimp' connector has a pre-installed fiber stub that the manufacturer has pre-polished. This stub has index matching gel on the inside end of the fiber (Figure 3-28). Because of this structure, the connector is actually a mechanical splice in a connector.

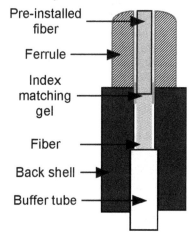

Figure 3-28: Internal Structure Of The Cleave And Crimp Connector

Installation by this method requires five steps:

- ➢ Cable end preparation
- ➢ Fiber end preparation
- ➢ Cleaving
- ➢ Insertion
- ➢ Crimping

The first two steps are the same as for all installation methods. The dimensions are different from those of other methods.

3.5.1 CLEAVING

The performance of this connector is strongly determined by the quality of the cleave on the fiber. Obtaining low and consistent cleave angles requires a high quality cleaver. Such a cleaver can cost more than $1200. The installer will save more than the additional cost of this cleaver from improved yield and reduced labor cost. To achieve the best results, the installer:

▶▶Uses a precision cleaver

Cleave and crimp connectors grip the cladding and the primary coating or buffer tube. A long cleave length may prevent the crimp from creating a proper grip through the primary coating and buffer tube. Without a proper grip, the connector can exhibit high loss or slippage of the fiber from the connector. A short cleave length may prevent the cleaved end from contacting the preinstalled fiber.

▶▶Control cleave length tightly

The installer cleaves the fiber to the cleave length and tolerance indicated in the connector instructions. A typical cleave length tolerance is ± 0.5 mm.

3.5.2 INSERTION

Contamination of the fiber end results in increased loss. Such contamination can block light from its proper path or create a gap between the fiber ends. There are two methods for avoiding such contamination.

▶▶Minimize delay

▶▶Avoid touching the cleaved end against anything

After cleaving the fiber, the installer inserts the fiber into the connector without delay. Delay can allow dirt and dust to collect on the end of the fiber, resulting in high loss.

During insertion of the fiber, the installer does not touch the cleaved end against anything. Doing so may contaminate the fiber end, resulting in high power loss. In addition, the fiber should not be placed on any surface. Such action results in contamination.

A common misconception is that the fiber should be cleaned after cleaving. Such cleaning is more likely to contaminate than to clean the cleaved end.

After inserting the fiber, the installer verifies that the fiber has not slipped from its fully inserted position. Such slippage creates a gap between the fiber and the preinstalled fiber. This gap results in high loss (Figure 2-2).

▶▶Verify full contact of fiber with internal fiber

3.5.3 CRIMPING

Crimping of a cleave and crimp connector is similar to that of any other connector. However, the crimper is unique to the product. Finally, there are two crimps: one to grip the cladding and the second to grip the fiber through the primary coating and buffer tube.

3.6 SUMMARY

The information, principles and methods presented in this chapter indicate the need for attention to detail. Such attention is necessary to address the subtleties of achieving low loss, low reflectance, high reliability, and low installation cost.

3.7 REVIEW QUESTIONS

1. Organize the following polishing films in the proper sequence of use: 3-μ, 12-μ, 0.5-μ, and 1-μ.

2. Is tap water the best source of liquid for wet polishing? Justify your answer.

3. Is cracking common in connectors using quick cure adhesive? Justify your answer.

4. Is wet polishing preferred for all polishing? Justify your answer.

5. Does oven curing of quick cure adhesive takes less than 2 minutes? Justify your answer.

6. Should the strength members be cut longer than necessary if a connector requires that the strength members be crimped to the back shell? Justify your answer.

7. Is undercutting is desirable? Justify your answer.

8. Should installation of a fiber into all connectors be done as slowly as possible to avoid breakage and other problems? Justify your answer.

9. What single action almost completely eliminates fiber breakage during insertion?

10. Your technician states that the cause of cracking in his room temperature cured connectors was the epoxy. Do you believe him? Justify your answer.

11. Your technician is terminating a 12 fiber premises cable with cleave and leave connectors. He has just finished cleaving all 12 fibers, which he has placed on a work mat. He is about to insert the first fiber into a connector. Is there anything wrong with this procedure? If so, what?

12. Your technician is polishing both multimode and singlemode connectors. To simplify his process, he plans to use the same polishing films. He plans to finish the polishing with a 1-μ film. Is there anything wrong with this procedure? If so, what?

13. Your technician states that the cause of high loss in his hot melt adhesive connectors was pistoning. Do you believe him? Justify your answer.

14. You are to make a decision on which cleaver to purchase for your cleave and crimp

connectors. You plan to install 1200 connectors. The inexpensive cleaver costs $300. The expensive cleaver costs $1200. The connectors cost $15. You have been led to believe that you will lose 5 % of the connectors with the expensive cleaver and 13 % with the inexpensive cleaver. Which cleaver should you purchase?

4   INSPECTION

Chapter Objective: you learn how to inspect and rate the appearance of a connector. You learn how to interpret the appearance of 'bad' connectors in order to identify the appropriate corrective action.

4.1   APPLICABILITY

This chapter applies to the inspection of all fiber optic connectors that require polishing. For all connectors except the 'cleave and crimp' products, the condition, or appearance, of the end face of the connector correlates highly with the power loss of the connector.

That is, a 'good' appearance of the end face correlates highly with low loss. Because the 'cleave and crimp' products include a mechanical splice in the back shell, the correlation between end face appearance and power loss is not as high as for connectors that require polishing.

4.2   EQUIPMENT

  ➢ A 400-magnification connector inspection microscope with an infrared filter[8]

  ➢ Lens grade tissues (Kim wipes or equivalent)

  ➢ 98% isopropyl alcohol

  ➢ Electro-Wash® Px

Some fiber optic professionals recommend magnifications of 100 and 200. These lower magnification microscopes allow the installer to miss features that increase loss and reflectance. The 400-magnification microscope enables installers to see all features of concern. Unfortunately, this magnification has two disadvantages. The first disadvantage is that it will reveal features that are irrelevant. Use of 400x microscopes requires the installer to learn those features that can be ignored.

The second disadvantage of 400x is inability to view the entire ferrule surface. An inspection at 400x may not reveal dirt on the ferrule outside of the field of view.

100x inspection will. In this case, loss test results will not be consistent with microscopic appearance. Cleaning of the ferrule is the solution to this inconsistency.

Note: photographs in the text of this chapter are at a magnification of slightly less than 200; photographs in the Review Questions are at various magnifications. To determine the actual magnification, multiply the fiber diameter, in inches, by 200.

4.3   PROCEDURE

4.3.1   CLEANING

Prior to inspection, all connectors require cleaning. The installer may perform this cleaning as part of the installation activity or just prior to inspection.

The installer performs cleaning with tissues and a cleaning solution. The tissues are lens grade and lint free, such as Kim Wipes or woven tissues.

The liquid can be 98% isopropyl alcohol or a connector cleaning solution such as Electro-Wash® Px (Chemtronics, Kennesaw GA). A third alternative is lens grade tissues pre-moistened with isopropyl alcohol. Note that medical alcohol pads and rubbing alcohol are not suitable for cleaning fiber optic connectors. Both may contain water and oil, which leave a residue.

This author's observation is that isopropyl alcohol is less expensive than Electro-Wash. Electro-Wash is more effective, more expensive, and easier to use than is isopropyl alcohol. This author uses both.

The installer wipes the connector twice: with a moistened tissue area and with a dry tissue area. The wet tissue can leave watermarks on the connector. Watermarks complicate interpretation of

---

[8] www.Westoverscientific com

the condition of the connector and can result in increased reflectance.

With ElectroWash Px, the installer creates a cleaning pad of tissue. He wipes from a moistened area to a dry area three times. With this method, ElectroWash leaves no residue.

4.3.2 VIEWING INSTRUCTIONS

The installer inspects with a microscope that has an adapter that will accept the connector ferrule. He performs four steps on each connector :

> ➢ Remove the cap from the connector

> ➢ Install the connector into the microscope

> ➢ Turn on and focus the microscope

> ➢ View and evaluate the connector (Figure 4-1)

Figure 4-1: Use Of Inspection Microscope

4.3.3 BACK LIGHT

To inspect with back lighting, another person holds a white light to the connector on the opposite end of the fiber. The core should light up. If the core does not light up, the fiber is broken. The installer inspects and evaluates the connector as in 4.4.

Whenever possible, the installer performs connector inspection both with (Figure 4-2) and without (Figure 4-3) back light. Backlight can both reveal and conceal features in the core. The principle is:

▶The installer rates the connector 'good' if it is 'good' both with and without backlight.

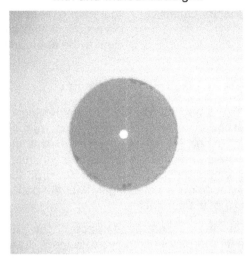

Figure 4-2: Connector With Back Light

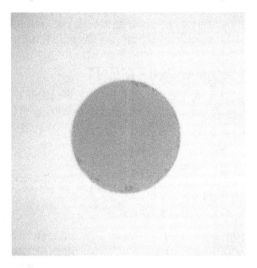

Figure 4-3: Same Connector No Back Light

If the cladding lights up during back lighting, the fiber is broken in the connector, usually in the ferrule (Figure 4-4 and Figure 4-5). While infrequent with adhesive connectors, backlight in the cladding is not unusual in 'cleave and crimp' connectors (3.5) installed by novices.

During back lighting, multimode cores can exhibit concentric rings. These rings result from the multiple layers in the core (1.2.1).

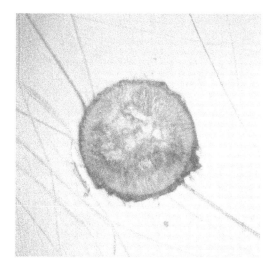

Figure 4-4: Shattered End With Back Light

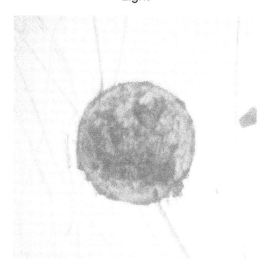

Figure 4-5: Shattered End

4.4   EVALUATION CRITERIA

In order to rate the condition of a connector, the installer need remember two words:

➢ Core

➢ Contact

The light travels within the core, or mode field. Therefore the condition of the core determines the loss of the connector. The connectors must make full contact in order to avoid excess loss from non-contact (Figure 2-2).

With these two words, the installer can understand the criteria for rating a connector as 'good'.

A good connector has a

▶'Good' core

▶Clean cladding

▶Clean ferrule

4.4.1   CORE

Any condition that can divert or block the light from its normal path will increase the loss of the connector. Thus, a 'good' connector must have a core that is (Figure 4-6):

▶Round

▶Clear

▶Featureless

▶Flush

A round core means that none of the core is below the surface of the ferrule (Figure 3-18). If the core is below this surface, it cannot be featureless, because it cannot be polished.

A clear core has no oil or grease that can make the image unclear or fuzzy. A featureless core has no features that can block or divert the light from its normal path. Features are

➢ Cleaning residue (Figure 4-7)

➢ Scratches (Figure 4-8)

➢ Cracks (Figure 4-9 and Figure 4-10)

➢ Pits

➢ Dirt (Figure 4-11)

➢ Dust

One final concept on features: if a feature, such as a speck of dirt, is large enough to be visible, it is too large to accept. If repeated cleaning does not remove it, it is not dirt, but a pit in the surface of the fiber.

Finally, the core must be flush with the ferrule surface. If it is above the surface, the fiber can be damaged from use. If it is below the surface, there is an air gap between this fiber and the mating fiber. As you learned in Chapter 2, air gaps increase loss.

It is possible to determine that the core is flush with the ferrule at 400 x. It may not be possible at lower magnifications. The core is flush with the ferrule when the fiber and the ferrule surface are both in focus at the same time.

4.4.2 EVERYWHERE ELSE

To achieve low loss by full core contact, the cladding and ferrule surface must be:

▶Clean

Dirt on ferrule(Figure 4-12) and cleaning residue (Figure 4-13) can prevent full ferrule contact.

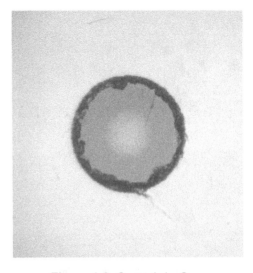

Figure 4-8: Scratch In Core

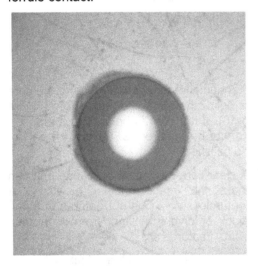

Figure 4-6: Good Connector- Back Light

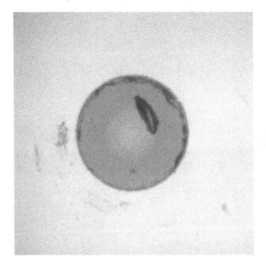

Figure 4-9: Crack And Non-Round Core

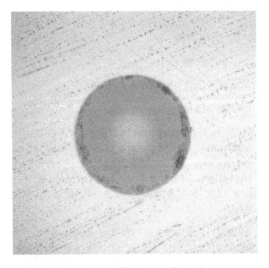

Figure 4-7: Cleaning Residue

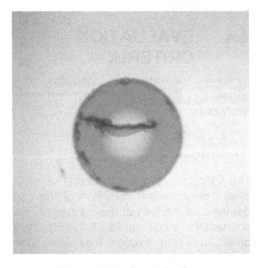

Figure 4-10: Crack In Core

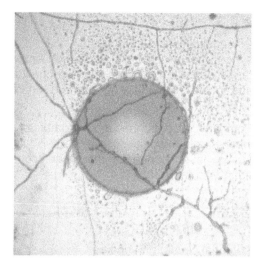

Figure 4-11: Dirt On Connector

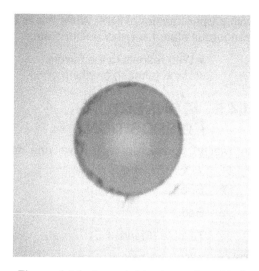

Figure 4-14: Acceptable, Imperfect Clad

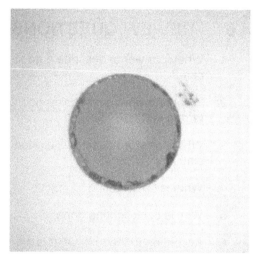

Figure 4-12: Dirt On Ferrule

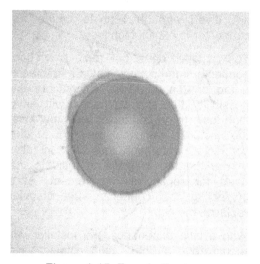

Figure 4-15: Ferrule Features

4.4.3 BAD CLAD? BE GLAD!

As light travels in the core, the cladding need not be perfect. An imperfect cladding means polishing was not perfect (Figure 4-8, Figure 4-12, Figure 4-14). The polishing procedure may need improvement. Thus, the cladding need not be round. It need not be completely present. It need not meet the criteria for a good core (4.4.1). However, it must be clean.

4.4.4 FERRULE FEATURES

The only requirement for the ferrule is that it be clean. Features in the ferrule surface (Figure 4-4, Figure 4-15) will not interfere

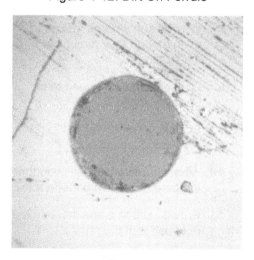

Figure 4-13: Cleaning Liquid Residue

with proper transfer of light. While this is a dangerous statement, this author states:

▶With respect to the ferrule surface (almost) anything goes.

4.4.5 CONNECTOR DISPOSITION

In reality, connectors will have one of three evaluations:

➢ Good

➢ Bad

➢ Terrible (Figure 4-5)

The definition of a 'good' connector is that of a *perfect* connector. Small scratches, cracks or contamination in the core may not divert enough light to cause the connector to exhibit high loss.

The installer does not replace a 'bad' connector immediately. Instead, he places a tag on the connector to indicate the possibility that the connector will have high loss. If the connector tests low loss, the installer removes the tag. If the connector tests high loss, the installer leaves the tag in place to indicate the need for repair or replacement. After inspecting the connector, the installer replaces the cap.

With a little experience, the installer will recognize terrible connectors. He replaces these connectors prior to testing.

4.5 TROUBLESHOOTING

4.5.1 DIRT ON CONNECTOR

Potential cause: dirt on microscope lens

Action: clean lens; rotate connector in microscope; if dirt does not move, dirt is on lens. Clean lens.

4.5.2 FAINT STAINS ON CONNECTOR

Potential cause: moisture stains from alcohol
Action: dry wipe immediately after cleaning with alcohol; or clean with Electro-Wash Px

4.5.3 NO FIBER FOUND

Potential cause: out of focus
Action: move focusing ring through entire range of motion; or, back light connector and move focusing ring through entire range of motion

Potential cause: connector not completely inserted into microscope
Action: insert connector into microscope adapter completely

4.5.4 CORE DOES NOT BACKLIGHT

Potential cause: broken fiber
Action: locate break with fault indicator or VFL

4.6 REVIEW QUESTIONS

1. What four words describe a good core?

2. What one word describes a good ferrule surface?

3. What one word describes a good cladding?

4. What is backlighting?

5. Why is back lighting done?

6. According to the text, what is the best magnification for inspecting connectors?

7. Why do some cores test good even though they may have a 'bad' rating?

8. Does the cladding need to be perfectly round?

9. Explain your answer to the previous question.

10. What characteristic tells you that the core is flush with the ferrule?

11. Provide two reasons that a fiber is not flush with the ferrule.

For each of the photographs in Review Questions 13 to 30, rate the connector as either 'good' or 'bad'. Then identify the reason for the connector being bad. Multiple reasons are possible. Use

- ➢ Not round
- ➢ Feature in core
- ➢ Dirt on core
- ➢ Dirt on cladding
- ➢ Dirt on ferrule
- ➢ Not flush

12. Rate this connector.

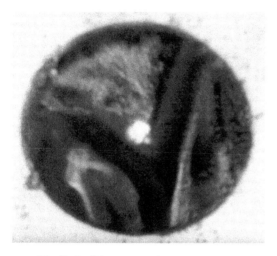

13. Rate this connector.

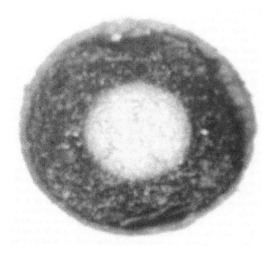

14. Rate this connector.

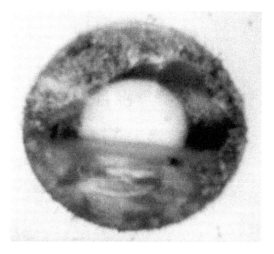

15. Rate this connector.

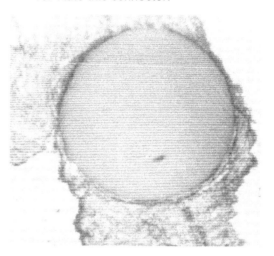

16. Rate this connector.

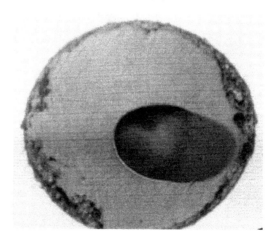

17. Rate this connector.

18. Rate this connector.

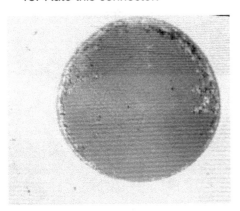

19. Rate this connector.

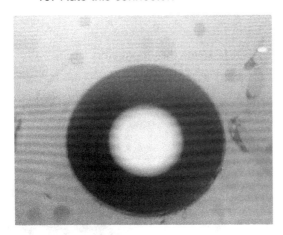

20. Rate this connector.

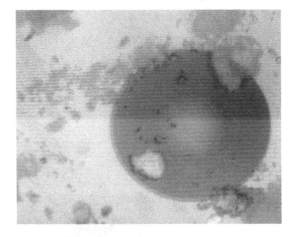

21. Rate this connector.

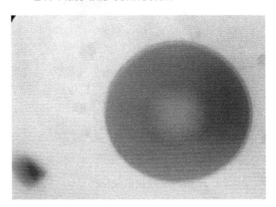

22. Rate this connector.

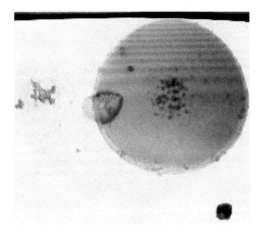

23. Rate this connector.

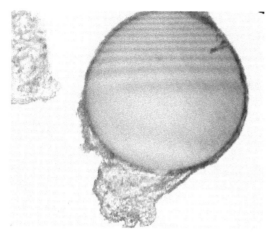

24. Rate this connector.

25. Rate this connector.

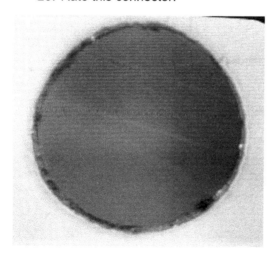

26. Rate this connector.

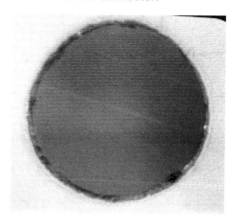

27. Rate this connector.

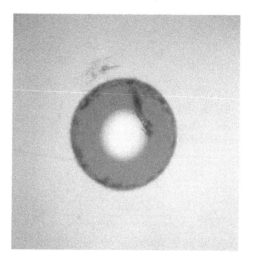

28. Rate this connector.

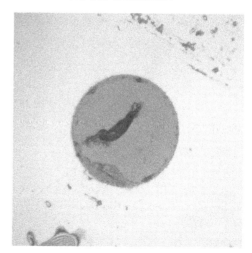

29. Rate this connector.

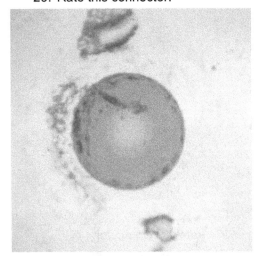

5 EPOXY INSTALLATION

Chapter Objectives: you install epoxy SC connectors with low loss and high reliability. In addition, you achieve singlemode reflectance below -50 dB.

5.1 INTRODUCTION

These instructions enable the novice installer to achieve low power loss and high reliability.

This chapter applies to installation onto five cable types. Three of the cables, single fiber, zipcord duplex and break out, have a 3 mm jacket on each fiber. Because of this jacket, we refer to these cables as having 'jacketed fibers'. The fourth type of cable, a premises cable, has a 900-µ tight buffer tube. Finally, these instructions enable installation onto loose tube cables with furcation kits on the fibers.

These instructions can apply to any connector, such as ST™-compatible and FC, using the appropriate dimensions and the appropriate crimp nest diameters.

The method of connector installation is epoxy. The epoxy requires heat curing for 10 minutes at 85° C. Another epoxy can be used with the appropriate cure time and temperature.

5.2 MATERIALS

The tools and supplies listed herein are required for installation of this connector by this method. We have chosen some of these tools and supplies on the basis of their superior performance. We have chosen others, on the basis of their price. Finally, we have chosen some of these tools and supplies based on their convenience.

Sources of equipment and supplies are: Fiber Optic Center Inc. (FOCI) 800-ISFIBER and Fiber Instrument Sales (FIS), 800-5000-FIS.

For installation of these connectors, the installer requires a connector installation tool kit, which includes the following items.

➢ Multimode or singlemode SC connectors with ceramic ferrules

➢ Crimper with nests appropriate to connector (Example: 0.137" and 0.190")

➢ Work mat (Clauss Fiber-Safe™)

➢ Tubing cutter (Ideal 45-162)

➢ Kevlar scissors (with ceramic blade)

➢ Miller buffer tube and primary coating stripper (FO-103-S) or Clauss NoNik stripper (NN203)

➢ Syringe (Fiber Optic Center Inc., FOCI)

➢ Epoxy (FIS part number H05-100-R2 or equivalent; FOCI part number AB9112)

➢ 85° C. connector curing oven (FIS part number F1-9772)

➢ Lens grade tissue (Kim-Wipes™ or equivalent)

➢ 98% isopropyl alcohol (FIS)

➢ Alternative to isopropyl alcohol: Alco pads or Opti-Prep Pads

➢ Connector cleaner, ElectroWashPx, (ITC Chemtronics)

➢ Wedge scriber (Corning Cable Systems Ruby Scribe, 3233304-01)

➢ Small plastic bottle (for bare fiber collection)

➢ Lens grade compressed gas (Stoner part number 94203 or equivalent)

➢ One SC polishing tool or one tool per film (FIS or FOCI)

➢ One hard rubber polishing pad, or one pad for each polishing film, except for the air polish film (FIS part number PP 575)

➢ 400 x connector inspection microscope with ST-compatible or SC fixture (Westover Scientific)

Multimode polishing requires these materials:

➢ 12-µ polishing film (FOCI part number AO12F913N100)

➢ 3-µ polishing film (FOCI part number AO3T913N100)

➢ 1-µ polishing film (FOCI part number AO1T913N100

➢ Optional: 0.5-µ polishing film (FOCI part number AO05T913N100)

Singlemode polishing requires these materials:

➢ 1-µ diamond polishing film (Fiber Optic Center Inc., part number D1KT403N1

➢ 0.5-µ diamond polishing film (Fiber Optic Center Inc., part number D05CF403N1

➢ 0.3-µ alumina polishing film (Fiber Optic Center Inc., part number D5BF403N1

➢ Polishing extender solution (Fiber Optic Center Inc.)

Scissors with ceramic blades provide the best results, but have the highest cost. Scissors with other types of blades will work. The polishing tool is also called a fixture and a puck. No pad is required for the 12-µ air polish film.

5.3   PROCEDURE

5.3.1   SET UP OVEN

The installer plugs in the oven. If the oven has a temperature adjustment, he sets the temperature to that appropriate for the epoxy. He allows the oven to heat up for at least 15 minutes or for the length of time indicated in the oven instruction sheet.

The instructions are for an epoxy that cures in 10 minutes at 85° C.

5.3.2   PREINSTALL CABLE

The installer installs the cable into the enclosure without permanently attaching the cable to the enclosure. He pulls the cable out of the enclosure to a work surface near the enclosure.

5.3.3   REMOVE JACKET

By the appropriate procedure, the installer removes the outer jacket of a tight tube cable, i.e., a premises cable or a breakout cable. At this step, the installer does not remove the jacket of 3 mm simplex or 3 mm zip cord duplex.

For a loose tube cable, the installer prepares the end according to the dimensions for the enclosure. He installs a furcation kit (1.3.2.2) onto the fibers for each buffer tube.

5.3.4   INSTALL BOOTS AND SLEEVES

5.3.4.1   PREMISES CABLE

The installer installs the 900-µ boot on each buffer tube, small end first. He pushes the boots approximately 12" away from end of buffer tubes. He marks the length of buffer tube to be removed. Note that Figure 5-1 is an example of such a length. The installer uses the length appropriate to the connector to be installed.

Strip this much bare fiber

0.688"

Figure 5-1: Template For Bare Fiber Length

5.3.4.2   JACKETED FIBER CABLES

The installer installs the boots on all fibers to be terminated. The cable enters the small end of the boot first. He pushes the boots approximately 12" away from end of cable.

The installer installs the crimp sleeves on all fibers to be terminated. The cable enters the small end of the crimp sleeve first. He pushes the crimp sleeve approximately 12" away from end of cable. He does not push the crimp sleeve under the boot.

5.3.4.3   ZIP CORD DUPLEX

If necessary, the installer cuts the thin web that connects the channels. The installer separates the two channels by pulling the channels apart. He installs boots and crimp sleeves as in 5.3.4.2.

5.3.5   PREPARE EPOXY

The installer twists a needle onto the barrel of a syringe until the needle resists additional twisting. He removes the plunger from the barrel.

He checks the expiration date of the epoxy. If the date has not passed, he uses the epoxy.

He mixes the two-part epoxy by removing the separator and rubbing the package with a mixing roller. As an alternative method, the installer rubs the package over a dull edge, such as the edge of a table. While mixing, he moves all of the epoxy components from one end of package to the other a minimum of 15 times or until the color of the mixture is uniform.

The installer squeezes all the epoxy into one end of the package. He squeezes the epoxy away from one corner of package. He cuts off 1/8" of the corner. He squeezes as much epoxy as possible onto the side of the inside of the barrel of the syringe. He does not allow the epoxy to run onto the needle at the end of the syringe.

He inserts the plunger into the barrel 1/8" -1/4". He points the needle up, allowing all of the epoxy to run onto plunger. He presses the plunger into the barrel until all the air passes through the needle. He pulls the plunger back about 1/8", so that epoxy does not 'weep' from the needle. He places the needle on a non-absorbent material, such as the epoxy package.

5.3.6   PREPARE END

The installer repeats this step for all fibers to be terminated at this location.

5.3.6.1   PREMISES CABLE

If the installer uses a NoNik Stripper, he cleans the stripper by holding the stripper open. He snaps each cap once (Figure 5-2).

Occasionally, this method the Clauss stripper does not work. In that case, the installer blows out the stripper with lens grade gas.

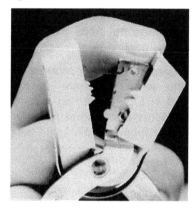

Figure 5-2: Cleaning Of Clauss Stripper

The installer grips the buffer tube firmly. The installer can grip by wrapping the cable around one finger 5 times (Figure 5-3) or by weaving the cable through his fingers.

The installer places the stripper on the buffer tube at the distance marked (Figure 5-1) or 1/2" from the end, whichever is shorter.

While holding the stripper at 90° to the fiber (

Figure 5-3), the installer pulls the stripper *in the direction of the arrow* on the stripper towards the fiber end slowly. He *allows* the buffer tube and primary coating to slide from the fiber. He does not force the buffer tube from the fiber.

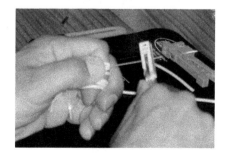

Figure 5-3: Fiber Straight In Stripper

If the installer uses a Miller Stripper, he cleans the stripper with lens grade gas. He grips the buffer tube firmly as described above.

The installer places the stripper on the buffer tube at the marked distance, or at 1/2" from the buffer tube end, whichever distance is less. He holds the stripper at 45° to the fiber. He pulls the stripper towards the end of the fiber slowly, allowing the buffer tube to slide from the fiber. He does not force the buffer tube and primary coating from the fiber.

If necessary, the installer repeats this step until the length of bare fiber is as indicated in Figure 5-1. He inspects the fiber to ensure that there is no buffer tube or primary coating remaining on the fiber.

> This instruction indicates a maximum strip length of 1/2". US-made cables can be stripped to at least this distance. Some US-made cables allow a longer strip length without breakage. For such cables, you may strip to that increased distance. Few cables allow more than 1.25 ".

5.3.6.2   JACKETED FIBERS

A 'jacketed' cable has a jacket on each fiber. Such a cable is simplex, zipcord duplex, or breakout (PFOI, 4).

5.3.6.2.1   REMOVE JACKET

The installer removes the jacket on each fiber by placing the tubing cutter at the indicated distance from the cable end (Figure 5-4 and Figure 5-5). He rotates the cutter around the cable once or twice. He removes the cutter and slides the jacket from the cable.

Note that Figure 5-4 is an example of such lengths. The installer uses the length appropriate to the connector to be installed.

Remove this much jacket

■■■■■■■■■■   1.25"

Leave this much buffer tube

■■■■■   0.625"

Leave this much aramid yarn

■■■   0.313"

Figure 5-4: Template Of Typical Strip Lengths

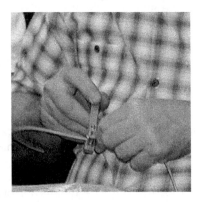

Figure 5-5: Tubing Cutter On Cable

5.3.6.2.2   STRIP BUFFER TUBE

The installer holds the aramid yarns back against the remaining jacket. He marks the buffer tube length to be left (Figure 5-4). He strips the buffer tube and primary coating as in 5.3.6.1.

5.3.6.2.3   TRIM STRENGTH MEMBERS

The installer separates the aramid yarn from the buffer tube. He twists and loops the yarn over the scissor blade. He cuts the yarn to the length in Figure 5-4 (Figure 5-6).

Some connectors allow trimming the strength member as the second step. For such connectors, the strength member length must be less than the buffer tube length.

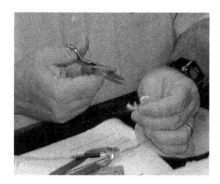

Figure 5-6: Cutting Strength Member

5.3.6.3   CABLE CONTINUITY TEST

This test is optional, but recommended. This test works on both multimode and singlemode fibers to several thousand feet.

The installer holds a high intensity light source close to the fiber ends at one end of the cable. A 1 W single, LED flashlight works well. An installer at the opposite end views the fibers. Each fiber core should glow or sparkle, indicating continuity. Note that continuity on a tight tube cable does not prove that there are no breaks.

5.3.7   CLEAN FIBER

The installer performs this step for each fiber immediately prior to inserting it into a connector. The installer moistens a lens grade tissue with isopropyl alcohol. He folds the moist area around the fiber (Figure 5-7). He pulls the fiber through the fold twice. He inspects the fiber for particles on the cladding. If necessary, he cleans the fiber repeatedly until the cladding is free of all particles.

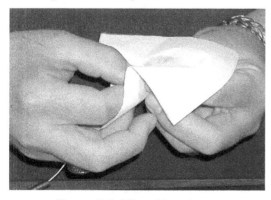

Figure 5-7: Fiber Cleaning

5.3.8   FIBER STRENGTH TEST

This step is optional. The installer performs this step to troubleshoot fiber breakage. The installer performs this step for each fiber just prior to inserting it into a connector.

The installer pushes each fiber against a lens grade tissue on the work surface. If the fiber bends without breaking, the installer has not damaged the cladding. If there is a single defect on the cladding, the fiber will break.

> ➤ SAFETY WARNING: the installer should not push a bare fiber against his finger. Doing so can result in the fiber penetrating the skin and breaking.

5.3.9   CONNECTOR TESTS

At this step, the installer can make two tests. Both tests are optional. Both these tests are troubleshooting steps for failure of fibers to fit into connectors or for fiber breakage. If necessary for troubleshooting or desirable for practice, the installer repeats these steps for all connectors.

5.3.9.1   WHITE LIGHT TEST

While aiming the ferrule towards a light source, such as a window or ceiling light, the installer looks into the back shell of a connector. He should see a sparkle that indicates the presence of a fiber hole and the absence of contamination in the connector. Such contamination could block the path of the fiber.

5.3.9.2   DRY FIT TEST

While resting his hands together, the installer feeds the fiber into central tube in the back shell of the connector (Figure 5-8). As soon as the fiber enters the back shell, he twists the connector back and forth. As long as the fiber continues feeding into the back shell without bending, he continues twisting the connector and inserting the fiber. If the fiber bends, he withdraws the fiber approximately 1/8", twists the connector back and forth and feeds the fiber into the back shell.

Figure 5-8: Central Tube Of SC Connector

The installer must see the fiber protruding beyond the tip of the ferrule. This protrusion indicates that the fiber and the hole are compatible, that the fiber is sufficiently long and that there is nothing blocking the fiber path through the ferrule. The installer removes the fiber from the connector and places that connector and fiber together.

Note: if the installer places the fiber on a surface, he must repeat 5.3.7 prior to inserting the fiber into the connector.

5.3.10 EPOXY INJECTION

Before each use, the installer wipes the epoxy from the outside of the needle. He places the needle into the central tube of the connector back shell (Figure 5-8) until the needle butts against the inside end of the ferrule. While maintaining pressure on the needle, he presses the plunger until the epoxy flows through the fiber hole in the end of the ferrule. As soon as the adhesive flows through the fiber hole, he removes the needle from the connector. He pulls the plunger 1/8" out of the barrel of the syringe to prevent 'weeping'. With experience, he can repeat this step for up to 12 connectors.

The installer can experiment with filling more than 12 connectors. However, all epoxies have a useful life, called the pot life. Should the epoxy harden excessively, the installer can experience two problems. First, the installer may not be able to feed the fiber through the ferrule. Second, the installer may break the fiber, resulting in loss of the connector.

This next step is optional and recommended for experienced installers.

With lens grade tissue, the installer wipes the epoxy from the tip of the ferrule.

Some installers will benefit from leaving the epoxy on the tip of the ferrule. This epoxy provides additional support for the fiber during polishing. However, this additional epoxy increases polishing time.

5.3.11 FIBER INSERTION

The installer inserts the fiber into the center tube in the back shell of a filled connector. As soon as the fiber enters the tube, he twists the connector back and forth. As long as the fiber continues into the tube without bending, he continues twisting the connector and inserting the fiber. If the fiber bends, the installer withdraws the fiber 1/8", twists the connector back and forth and feeds the fiber into the back shell.

When the buffer tube stops further motion of the fiber into the ferrule, he inspects the tip of the ferrule. He should see the fiber protruding beyond the tip of the ferrule. As long as the amount of protrusion is not excessive, it is not critical.

If he does not see the fiber protruding beyond the tip of the ferrule, he removes the fiber. The installer examines the fiber. If the fiber is shorter than it was after preparation, the broken fiber is inside the connector. He discards the connector. If the fiber has not broken, the installer can reuse the same connector.

If the fiber was short, he repeats end preparation (5.3.6). If the time to prepare a new end is short, the installer can insert the new end into the same connector.

If the fiber is jacketed fiber design, the installer proceeds to 5.3.12. If the cable is a premises or loose tube design, the installer proceeds to 5.3.13.2.

5.3.12 CRIMP SLEEVE

He slides the crimp sleeve up to and over the back shell of the connector. While sliding the sleeve over the back shell, he rotates the sleeve back and forth. This distributes the strength members uniformly around the back shell.

He crimps the crimp sleeve with the crimp nest or nests specified by the connector manufacturer (Figure 5-9). He crimps the large diameter of the sleeve over the back shell of the connector and the small diameter of the crimp sleeve over the jacket.

Figure 5-9: Crimping Crimp Sleeve

The installer repeats this step for each connector that has epoxy and has fiber protruding beyond the tip of the ferrule. He checks the length of the strength members. They should be flush with the end of the crimp sleeve or less than 1/16" longer than the crimp sleeve.

> Caution: prior to crimping, check the position of the crimp sleeve: it should be in a nest, not between nests.

5.3.13 CONNECTOR INSERTION

5.3.13.1 JACKETED FIBERS
While holding onto the back shell of the connector, the installer centers the connector over the oven port (Figure 5-10). Slowly, he lowers the connector into the port. If he feels any 'springiness' or resistance, the fiber is off center and touching the inside of the oven. If he detects such resistance, he lifts the connector 1/4", re-centers the connector in the oven port and reinserts slowly. If the springiness persists, the fiber protruding beyond the ferrule is excessively long. The installer trims the excess length without breaking the fiber below the end of the ferrule.

> Caution: this step is difficult.

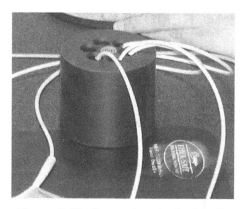

Figure 5-10: Connectors In Curing Oven

5.3.13.2 PREMISES CABLES
The installer slides the boot over the back shell. While squeezing the boot against the back shell of the connector, the installer inserts the connector into the oven as in 5.3.13.1. He does so without placing tension on the fiber. He records the cure start time for each connector or batch of connectors.

> With no crimp sleeve on the connector, tension can cause the fiber to withdraw into the ferrule.

After the specified minimum cure time, he removes each connector from the oven. The cure time for the epoxy in this procedure is ten minutes.

If no oven position is available for a new connector, the installer places any uncured connector horizontally on the work surface in such a manner to avoid tension on the fiber.

The installer may tape the cables or buffer tubes to the work surface so that the buffer tubes do not slide out of the connectors. Should the buffer tubes or cables slide, the fibers protruding beyond the end of the ferrule may break.

> Do not pause or delay installing fibers into filled connectors because the epoxy hardens continuously after being mixed.
>
> You may cure the epoxy for more than the minimum curing time, as excessive cure time causes no problems.

Repeat this step until you have cured and allowed all connectors to cool. Do not force cool connectors.

5.3.14 REMOVE EXCESS FIBER

While watching the fiber at the end of the ferrule, the installer pulls gently on the jacket or buffer tube. The fiber should not move. If it moves, the epoxy has not cured. Once the epoxy has cured, the installer proceeds.

While resting his hands together, the installer places the wedge surface of a scriber onto the end of the ferrule (Figure 5-11). He moves the blade of the scriber up to the fiber so that the scriber gently touches the fiber at the top of the epoxy bead. The fiber should not move or bend.

The installer moves the scriber back and forth along the fiber 1/16" once. He should not break the fiber with the scriber. He should not 'saw' the fiber with the scriber.

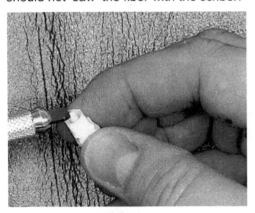

Figure 5-11: Scribing Excess Fiber

The installer holds the connector with the fiber pointing up. He grips the ferrule with his thumb and forefinger lightly. He slides his thumb and forefinger up the ferrule towards and over the fiber. He grips and pulls the fiber away from the tip of the connector. The fiber should break easily. He places of the fiber in the fiber collection bottle.

Note: the installer does not try to grasp the fiber from the side. Doing so can break the fiber in bending, not in tension. If the fiber breaks in bending, it may break below the surface of the connector. With ceramic ferrules and standard polishing films, polishing does not remove this condition.

If the fiber does not break, the installer scribed epoxy on the fiber and not the fiber. He repeats this step until the fiber breaks.

5.3.15 AIR POLISH

The installer holds the connector with the ferrule pointing up. He holds a 12-µ polishing film at one edge with the dull (abrasive) side down (Figure 5-12). He places the opposite edge of the film above the connector. He curls the front and back edges of the film down. With a light pressure, he rubs the film against connector until the fiber is flush with the bead of epoxy. If he scribed the fiber just above the surface of the epoxy bead, this step will take less than 10 seconds. The installer does not remove all the epoxy with this film.

The installer checks the fiber by bringing a finger down onto the tip of the ferrule from the top. If he cannot feel the fiber, the fiber is flush, or nearly flush, with the epoxy.

Caution: do not rub a finger *across* the tip of the ferrule. If the fiber is not flush with the adhesive, the installer may snag and break the fiber. In addition, the broken fiber may pierce the installer's skin.

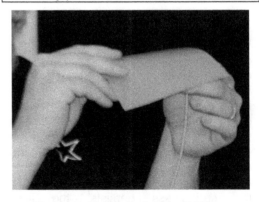

Figure 5-12: Air Polishing Of Fiber

5.3.16 MULTIMODE POLISH

For single connectors, the installer follows the procedure in this section for that connector. For multiple connectors, the installer can perform each step on all connectors prior to performing the next step; i.e., a 'batch process'.

5.3.16.1   CLEAN CONNECTORS

The installer moistens a lens grade tissue with isopropyl alcohol. He wipes the ferrules of all connectors with this moistened tissue. This removes 12-µ grit.

5.3.16.2   CLEAN EQUIPMENT

With lens-grade compressed gas or isopropyl alcohol, he cleans off the top surface of the polishing pad(s), both sides of the polishing film(s), and the polishing tool. He places the polishing film(s), dull side up, on the pad(s).

5.3.16.3   FIRST POLISH

The purpose of the first polish is removal of the epoxy. Additional time after such removal does not improve the loss, but can result in increased loss.

The installer places the polishing tool, hereafter 'tool', on the 3-µ film. He inserts the connector into the polishing tool.

Light pressure means that the installer holds the connector against the film without compressing the spring inside the connector. While holding both the connector and the tool (Figure 5-13), he moves the tool with a light pressure in a ½" high, figure-8 pattern slowly. If he feels scratchiness, the fiber is not flush with the epoxy. He maintains a light pressure and a slow, ½" high figure-8 pattern until the scratchiness ceases.

Figure 5-13: Polishing

After the scratchiness ceases, the installer increases the size of the figure-8 pattern to cover the entire film. He may increase the speed of polishing. He does not increase the pressure. He continues polishing until the epoxy is completely removed.

The installer may detect a change in the friction of the connector on the film. This change indicates the removal of the last of the epoxy. If he does not detect this change, he removes the connector from the tool periodically. He views the tip of the ferrule by reflecting light off the tip. When the epoxy is completely removed, the tip of the ferrule will be glass smooth and shiny. The tip will have the same appearance as that of the side of the ferrule.

5.3.16.4   CLEANING

With lens-grade compressed gas or isopropyl alcohol and lens grade tissues, the installer cleans the connector and the tool.

5.3.16.5   SECOND POLISH

The purpose of the second polish is creation of a 'featureless,' low loss core (4.4). 'High pressure' for polishing means holding the connector in the polishing tool and pushing the inner housing down until the internal spring is fully compressed. The installer need not increase the pressure beyond the level at which the spring is fully compressed.

The installer places the polishing tool on the 1-µ film and applies high pressure to the connector for ten, large figure-8 patterns. Large means using the full area of the film.

5.3.16.6   THIRD POLISH

Some connector manufacturers recommend a third polish. This polish is optional This author has found no reduction in loss from this polish on multimode connectors. At best, this author has found only minor improvement in the microscopic appearance of the core.

The installer cleans the connector and polishing tool (5.3.16.4). The installer places the polishing tool on the 0.5-µ film. He inserts the connector in the tool and applies high pressure to the connector for ten, large figure-8 patterns.

5.3.17 FINAL CLEANING

The installer slides the boot over the back shell. He cleans the connector with one of the following methods.

5.3.17.1 BEST METHOD

The installer moistens a lens grade, lint free tissue with isopropyl alcohol. He wipes the sides and tip of the ferrule with this tissue. He sprays a one-inch diameter area of connector cleaner, such as ElectroWash Px, onto a small pad of lens grade tissues. Three times, he wipes the tip of the ferrule from the wet area to the dry area.

Optional: the installer blows out the cap with lens grade gas.

He installs the cap.

5.3.17.2 METHOD B

The installer moistens a tissue with isopropyl alcohol. He wipes the sides and the tip of the ferrule. Immediately, he wipes the tip of the ferrule with a dry tissue.

Optional: the installer blows out the cap with lens grade gas.

He installs the cap.

5.3.17.3 METHOD C

The installer wipes the sides and the tip of the ferrule with a pre-moistened lens grade tissue (Alco Pad or OptiPrep Pad). Immediately, he wipes the tip of the ferrule with a dry tissue.

Optional: with lens grade gas, the installer blows out the connector cap.

The installer installs the cap.

Do not use lens grade gas for final connector cleaning. This gas can leave watermarks that can increase loss and reflectance (Figure 4-7 and Figure 4-13).

5.3.18 INSPECTION

After polishing each connector, the installer inspects and rates the connector as in 4.4. In addition, he inspects the cladding for uniformity of appearance (Figure 4-6).

If the connector is not perfect, he tags the connector. If the connector is 'terrible' (4.4.5), he replaces the connector.

Note: this author does not recommend polishing more than one connector prior to microscopic inspection. Should the film be contaminated, cores can become scratched. Discovering such contamination after one connector is better than after a dozen!

5.3.19 WHITE LIGHT TEST

The installer performs a white light, continuity test as in 5.3.6.3 (Figure 5-14). The installer does not install connectors on the opposite end until all of the fibers have passed this continuity test. If one or more connectors fail this test, he troubleshoots these connectors to identify the problem. An OTDR test may be required.

The installer replaces the caps on all connectors that pass this test. He repeats this continuity test after he installs all connectors on the opposite end of the cable.

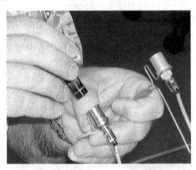

Figure 5-14: White Light Test

5.3.20 FINAL ASSEMBLY

The installer aligns the flat surfaces of the boot to the flat surfaces of the inner housing. He slides the boot over the back shell of the connector. The installer tugs on the cable or buffer tube gently. If the boot fits tightly, this tug straightens out the jacket or buffer tube inside the boot.

The top of the inner housing has missing corners (Figure 5-15). The top of the outer housing has the key. He holds the top of the inner housing and the top of the outer housing up. He inserts the inner housing into the end of the outer housing that is opposite the end with the key (Figure 5-16). He wiggles the inner housing until it snaps through the outer housing (Figure 5-17).

Figure 5-15: Alignment Housings

Figure 5-16: Insertion Of Inner Housing

Figure 5-17: Inner Housing Fully Inserted

5.3.21 SINGLEMODE POLISHING

Singlemode polishing can be used for initial installation or for restoration of low reflectance. Because of high cost, field polishing of singlemode connectors by hand is done rarely. Pigtail splicing is preferred.

We provide this procedure, with the recommendation that it be used for initial installation when no other option is available. This procedure can be used for restoring a singlemode connector to low reflectance by replacing the 3-μ film with a diamond film of 3-μ.

This author's assumption is that the defects causing high reflectance are shallow. Although deep defects can be removed with 5-μ diamond film, we recommend connector replacement, as polishing cost can be excessive.

5.3.21.1 CLEANING

As in 5.3.16.2, the installer cleans the polishing pads, the polishing tool and the following films:

> 3-μ

> 1-μ diamond

> 0.5-μ diamond

> 0.3-μ

5.3.21.2 FIRST POLISH

With the 3μ film, the installer polishes the connector as in 5.3.16.3.

5.3.21.3 CLEANING

With isopropyl alcohol and lens grade tissues, the installer cleans the connector. He scrubs the tool with lens grade tissues moistened with isopropyl alcohol at least twice.

5.3.21.4 SECOND POLISH

The installer places the polishing tool on the 1-μ diamond film. 'High pressure' for polishing means holding the connector in the polishing tool and pushing down the inner housing until the internal spring is fully compressed. The installer need not increase the pressure beyond the level at which the spring is fully compressed. The installer applies high pressure to the connector for 20 to 40, large figure-8 patterns. Large means using the full area of the polishing film.

The installer polishes for 20 strokes for new diamond film. He polishes 40 strokes after he has polished five connectors on the film. He can use the same film for at least 10 connectors.

5.3.21.5 CLEANING

The installer cleans the connector and tool as in 5.3.21.3.

5.3.21.6 THIRD POLISH

With a high pressure, the installer polishes the connector for 20-40 large figure 8 strokes on the 0.5-μ film. The installer polishes for 20 strokes for new diamond film. He polishes 40 strokes after he has polished five connectors on the film. He can use the same film for at least 10 connectors.

5.3.21.7 CLEANING

The installer cleans the connector and polishing tool as in 5.3.21.3. There must be no residue on the tool from previous polishing steps.

If the installer uses a single polishing tool for all films, he scrubs the tool with a toothbrush and isopropyl alcohol before each use on the 0.3-μ film. Failure to do so will degrade the condition of the core to worse than that after the 0.5-μ film.

An alternative polishing method is use of a separate tool for each film. This method reduces the time spent on cleaning, as only the connector need be cleaned.

5.3.21.8 FOURTH POLISH

The installer places the 0.3-μ film on a polishing pad. The installer adds two drops of the polishing extender liquid to the film. With a high pressure, he polishes the connector for 20 large figure 8 strokes. He replaces this film after five connectors.

5.3.22 FINAL CLEANING

As in 5.3.17, the installer cleans the connector with isopropyl alcohol and lens grade tissue three times. He installs a cap.

5.3.23 INSPECTION

Inspect as in 5.3.18.

5.4   TROUBLESHOOTING

5.4.1   INSTALLATION

5.4.1.1   FIBER BREAKAGE DURING INSERTION

Potential cause: dirt or primary coating on fiber
Action: clean fiber

Potential cause: debris in connector
Action: perform white light check of connector

If the connector fails this test, the installer blows out connector with lens grade gas.

If the connector still fails this test, he flushes the connector with isopropyl alcohol injected through the connector with a syringe and needle. If the connector fails after this flush, he replaces the connector.

5.4.1.2   FIBER BREAKS OR DOES NOT FIT INTO CONNECTOR

Potential cause: primary coating not completely removed
Action: restrip fiber

Potential cause: epoxy has partially hardened
Action: reduce number of filled connectors; reduce the time between epoxy injection and fiber insertion

5.4.1.3   EPOXY FLOWS FROM BACK SHELL

Potential cause: excessive adhesive in back shell
Action: discard the connector; after epoxy flows from fiber hole in ferrule, immediately withdraw needle from connector

5.4.1.4   EPOXY ON OUTSIDE OF BACK SHELL

Potential cause: failure to wipe needle before each use
Action: wipe epoxy from outside of needle before each use

5.4.1.5   EPOXY ON OUTSIDE OF INNER TUBE

Potential cause: failure to wipe needle before each use
Action: discard connector; wipe epoxy from outside of needle before each use

See 5.3.10

5.4.2 POLISHING

5.4.2.1   NON-ROUND CORE

Potential cause: fiber broken during first polishing due to incomplete air polishing
Action: replace connector and perform complete air polish

Potential cause: excessive pressure during first pad polishing

In this case, polishing sheared of the epoxy bead. This bead may be found on the first film.
Action: reduce polishing pressure

5.4.2.2 EXCESSIVE POLISHING TIME

Potential cause: large bead of epoxy due to failure to remove epoxy from tip of ferrule prior to insertion of fiber
Action: remove all epoxy from tip of ferrule after injecting epoxy

Potential cause: film worn out
Action: replace film

Potential cause: insufficient polishing pressure
Action: increase polishing pressure

5.4.2.3 NO APPARENT REDUCTION OF BEAD SIZE

Potential cause: dirt in ferrule hole of polishing tool
Action: clean hole in polishing tool with lens grade air or with pipe cleaner dipped in isopropyl alcohol.

5.4.2.4 FEW CORE AND CLADDING SCRATCHES

Potential cause: contamination of film by environment.
Action: replace film

Note: during field installation, contamination of polishing films is common. Often, such contamination is unavoidable. The installer cleans the film only if no replacement film is available. In addition, he can move the polishing location away from air vents and any other source of airborne dust.

5.4.2.5 EPOXY SMEARS ON FIBER DURING POLISHING

Potential cause: uncured or incompletely cured epoxy due to insufficient time
Action: increase curing time. Monitor dwell time in oven with written log

Potential cause: low curing temperature
Action: check oven temperature

Potential cause: epoxy used past its expiration date
Action: discard epoxy

Potential cause: epoxy allowed to freeze
Action: discard epoxy[9]

5.4.2.6 CRACKED FIBER

Potential cause: excessive pressure during scribing
Action: scratch fiber lightly during scribing

Potential cause: excessive temperature during curing
Action: repair or replace oven

5.4.3 SINGLEMODE POLISHING

5.4.3.1 PITS REMAIN AFTER FIRST 20 STROKES ON NEW 1-μ FILM

Potential cause: insufficient polish pressure
Action: increase pressure to that described in procedure

5.4.3.2 PITS REMAIN AFTER FIRST 20 STROKES ON 'OLD' 5-μ FILM.

Potential cause: film worn out
Action: replace film

5.4.3.3 HIGH REFLECTANCE WITH FEATURES NEAR CORE

Potential cause: features create roughness
Action: repolish starting with 3-μ diamond film

5.4.3.4 HIGH REFLECTANCE WITH NO FEATURES

Potential cause: dirt on cladding or ferrule outside of field of view
Action: clean ferrule and retest

Potential cause: roughness not visible at 400 x.
Action: repolish starting with 1-μ diamond film

5.4.3.5 HIGH REFLECTANCE

Potential cause: dirt on connector outside of field of view

---

[9] Some epoxy loses its ability to cure after it has been frozen. Store such epoxy so that it does not freeze.

Action: re-clean and retest

5.4.3.6 PITS APPEAR AFTER POLISHING ON 0.3-µ FILM[10]

Potential cause: the polishing tool was contaminated with debris from the prior polishing films
Action: thoroughly clean the tool prior to polishing on 0.3-µ film

5.5 SUMMARY

This summary is in three parts, one each for connector installation, multimode polishing, and singlemode polishing.

5.5.1 INSTALLATION

Set up oven. Plug in. Allow to preheat.

Install cable through enclosure.

Pull cable through enclosure to work surface.

Remove jacket, Kevlar and central strength member to proper lengths.

Mix epoxy thoroughly and fill syringe.

Install boots on all buffer tubes.

Remove buffer tubes with stripper to proper lengths.

Optional Step: give all connectors a white light test.

Optional Step: give all connectors a dry fit.

For up to 12 connectors, inject epoxy through fiber hole and a small amount into back shell.

Optional Step: Wipe all epoxy off tip of ferrules.

Clean fiber with lens grade tissue and isopropyl alcohol.

While rotating fiber or connector, insert fiber into back shell until fiber bottoms out against ferrule. Do not allow fiber to bend.

For jacketed fiber, slide crimp sleeve over back shell and crimp sleeve. For premises cable, bring boot over back shell.

While keeping connector centered over oven port, insert connector into oven

without breaking fiber. Allow to cure 10 minutes.

Remove connector from oven and allow to cool. Do not force cool.

Scribe and remove excess fiber.

Air polish excess fiber flush with epoxy.

Proceed to multimode or singlemode polishing procedure.

5.5.2 MULTIMODE POLISH

Clean all connectors.

Clean pads, films and tool. Place 3-µ film on pad and tool on film.

Install connector into tool.

With light pressure, polish in small figure 8 pattern until scratchiness ceases.

When scratchiness ceases, polish in a large figure 8 pattern until all the epoxy is gone. Polish the connector using the entire area of the film.

Clean the 1-µ film, the tool and the connector.

Install the connector into the tool. With high pressure, make 10 large figure 8 motions.

Clean the side and tip of the connector.

Proceed to connector inspection procedure (4).

Perform a white light continuity test after you install connectors on the first end of the cable.

Perform a white light continuity test after you install connectors on the second end of the cable.

Install boot. Tug jacket or buffer tube gently.

Install outer housing.

Clean and install cap on ferrule.

5.5.3 SINGLEMODE POLISH

Scribe and air polish the fiber flush with the epoxy.

Clean the ferrules of all connectors.

Clean off the polishing pad, the polishing films and the polishing tool.

---

[10] See Figure 4-14.

Polish on the 3-μ film to remove the epoxy.

With lens grade gas, clean the connector, pad, tool and the 1-μ diamond film.

With a heavy pressure, polish the connector on the 1-μ diamond film for 20-40 large figure 8 strokes.

With lens grade gas, clean the connector, pad, tool and the 0.5-μ diamond film.

With a heavy pressure, polish the connector on the 0.5-μ film for 20-40 large figure 8 strokes.

With lens grade gas, clean the connector, pad, tool and the 0.3-μ film thoroughly.

Place two or three drops of the polishing liquid on the film. With a heavy pressure, polish the connector on the 0.3-μ film for 20 large figure 8 strokes.

Clean the connector with alcohol and lens grade tissue three times.

Proceed to connector inspection (4).

Perform a white light continuity test after you install connectors on the first end of the cable.

Perform a white light continuity test after you install connectors on the second end.

Install boot. Tug jacket or buffer tube gently.

Install outer housing.

Clean and install cap on ferrule.

Install boot on jacketed cable.

6 QUICK CURE ADHESIVE INSTALLATION

Chapter Objectives: you install SC connectors with quick cure adhesive with low loss and high reliability.

6.1 INTRODUCTION

This chapter applies to the installation of a multimode SC connector with a two-part quick cure adhesive. The procedure for polishing of singlemode connectors installed with this adhesive system is in 5.3.21.

This chapter applies to installation onto five cable types. Three of the cables, single fiber, zip cord duplex and break out, have a 3 mm jacket on each fiber. Because of this jacket, we refer to these cables as having 'jacketed fibers'. The fourth type of cable, a premises cable, has a 900-µ tight buffer tube. Finally, these instructions enable installation onto loose tube cables with furcation kits on the fibers.

Quick cure adhesives do not require power to be cured. Such adhesives cure rapidly, resulting in relatively short installation time.

However, the adhesive creates a very small bead on the tip of the ferrule. This small bead provides limited support for the fiber during polishing. This limited support increases the difficulty of polishing.

In addition, quick cure adhesives may harden prior to full insertion of the fiber. In this situation, there is bare fiber inside the connector, a condition of reduced reliability.

6.2 MATERIALS AND SUPPLIES

The tools and supplies listed herein are required for installation of this connector by this method. We have chosen some of these tools and supplies on the basis of their superior performance. We have chosen others, on the basis of their price. Finally, we have chosen some of these tools and supplies based on their convenience.

Sources of equipment and supplies are: Fiber Optic Center Inc. (FOCI) 800-ISFIBER and Fiber Instrument Sales (FIS), 800-5000-FIS.

For installation of these connectors, the installer requires a connector installation kit, which includes the following:

➢ Multimode or singlemode SC connectors with ceramic ferrules

➢ Crimper with nests appropriate to connector (Example: 0.137" and 0.190")

➢ Work mat (Clauss Fiber-Safe™)

➢ Tubing cutter (Ideal 45-162)

➢ Kevlar scissors (with ceramic blades)

➢ Miller buffer tube and primary coating stripper (FO-103-S) or Clauss NoNik stripper (NN203)

➢ Syringe (Fiber Optic Center Inc., FOCI)

➢ Quick cure adhesive, Loctite part number 680

➢ Quick cure primer, Loctite part number 7649

➢ Lens grade tissue (Kim-Wipes™ or equivalent)

➢ 98% isopropyl alcohol

➢ Alternative to isopropyl alcohol: Alco pads or Opti-Prep Pads

➢ Connector cleaner (Electro WashPx, from ITC Chemtronics)

➢ Wedge scriber (Corning Cable Systems Ruby Scribe, 3233304-01)

➢ Small plastic bottle for bare fiber collection

➢ Lens grade compressed gas (Stoner part number 94203 or equivalent)

➢ SC polishing tool (FIS or FOCI)

➢ Two or more hard rubber polishing pads, one pad for each polishing film (FIS part number PP 575)

➢ 400 x connector inspection microscope with ST-compatible or SC fixture (Westover Scientific)

➢ 12-μ polishing film (FOCI part number AO12F913N100)

➢ 3-μ polishing film (FOCI part number AO3T913N100)

➢ 1-μ polishing film (FOCI part number AO1T913N100

➢ Optional: 0.5-μ polishing film (FOCI part number AO05T913N100)

You may use different quick cure adhesives, as long as you follow the instructions for those adhesives. Scissors with ceramic blades provide the best results, but have the highest cost. Scissors with other types of blades will work. The polishing tool is also called a fixture and a puck. No pad is required for the 12-μ air polish film. The listed films are for multimode polishing.

6.3   PROCEDURE

6.3.1   PREINSTALL CABLE

The installer installs the cable into the enclosure without permanently attaching the cable to the enclosure. He pulls the cable out of the enclosure to a work surface near the enclosure.

6.3.2   REMOVE JACKET

By the appropriate procedure, the installer removes the outer jacket of a premises cable or of a breakout cable. At this step, the installer does not remove the jacket of 3 mm simplex or 3 mm zip cord duplex.

6.3.3   INSTALL BOOTS AND SLEEVES

6.3.3.1   PREMISES CABLE

The installer installs the 900-μ boot on each buffer tube, small end first. He pushes the boots approximately 12" away from end of buffer tubes. He marks the length of buffer tube to be removed. Note that Figure 6-1 is an example of such a length. The installer uses the length appropriate to the connector to be installed.

Strip this much bare fiber

━━━━━ 0.688"

Figure 6-1: Template For Length Bare Fiber

6.3.3.2   JACKETED FIBER CABLES

The installer installs the boots on all fibers to be terminated. The cable enters the small end of the boot first. He pushes the boots approximately 12" away from end of cable.

The installer installs the crimp sleeves on all fibers to be terminated. The cable enters the small end of the crimp sleeve first. He pushes the crimp sleeve approximately 12" away from end of cable. He does not push the crimp sleeve under the boot.

6.3.3.3   ZIP CORD DUPLEX

The installer separates the two channels by pulling the channels apart. If necessary, the installer can cut the thin web that connects the channels. He installs a boot and crimp sleeve as in 6.3.3.2.

6.3.4   PREPARE ADHESIVE

The installer twists a needle onto the barrel of a syringe until the needle resists additional twisting. He removes the plunger from the barrel.

He checks the expiration date of the adhesive and primer. If the date has not passed, he uses both.

He squeezes the adhesive bottle so that the adhesive runs onto the side of the inside of the barrel without running onto the needle. He needs no more than 1-2 cc of adhesive.

He inserts the plunger into the barrel 1/8" -1/4". He points the needle up, allowing all of the adhesive to run onto the plunger. He presses the plunger into the barrel until all the air passes through the needle. He pulls the plunger back about 1/8", so that adhesive does not 'weep' from the needle.

> A trick for training: quick cure adhesives can be used past their expiration date. The curing time increases as the adhesive ages. Such increase a) allows the trainee increased time for insertion of the fiber and b) reduces the frequency of premature hardening. This author has used adhesives for up to three years past their expiration date. Eventually, the adhesive ceases to cure. Since this is so, the trainer must test expired adhesive prior to use.

6.3.5   PREPARE END

The installer repeats this step for all fibers to be terminated at this location.

6.3.5.1   PREMISES CABLE

If the installer uses a No Nik Stripper, he cleans the stripper by holding the head open and snapping each cap once (Figure 6-2).

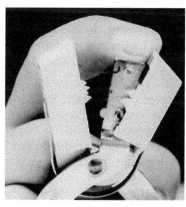

Figure 6-2: Cleaning Clauss Stripper

Occasionally, this method of cleaning does not work. In that case, the installer may blow out the stripper with lens grade gas.

The installer grips the cable or buffer tube firmly. The installer can grip the cable by wrapping the cable around one finger 5 times (Figure 6-3) or by weaving the cable through his fingers.

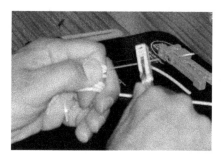

Figure 6-3: Fiber Straight In Stripper

The installer places the stripper on the buffer tube at the marked distance (Figure 6-1) or 1/2" from the end of the buffer tube, whichever distance is less.

While holding the stripper at 90° to the fiber (Figure 6-3), the installer pulls the stripper *in the direction of the arrow* on the stripper towards the fiber end slowly. He *allows* the buffer tube and primary coating to slide from the fiber. He does not force the buffer tube from the fiber.

If the installer uses a Miller Stripper, he cleans the stripper with lens grade gas. He grips the buffer tube firmly as described above. The installer places the stripper on the buffer tube at the marked distance or at 1/2" from the buffer tube end, whichever distance is less. He holds the stripper at 45° to the fiber. He pulls the stripper slowly towards the end of the fiber, *allowing* the buffer tube to slide from the fiber. He does not force the buffer tube and primary coating from the fiber.

If necessary, the installer repeats this step until the length of bare fiber is as indicated in Figure 6-1. He inspects the fiber to ensure that there is no buffer tube or primary coating remaining on the fiber.

> This instruction indicates a maximum strip length of 1/2". US-made cables can be stripped to at least this distance. Some US-made cables allow a longer strip length without breakage. For such cables, you may strip to that increased distance. Few cables allow more than 1.25 ".

6.3.5.2   JACKETED FIBER CABLES

6.3.5.2.1   REMOVE JACKET

The installer removes the inner jacket by placing the tubing cutter at the indicated distance from the cable end (Figure 6-4 and Figure 6-5). He rotates the cutter around the cable once or twice. He removes the cutter and slides the jacket from the cable.

Note that Figure 6-4 is an example of such lengths. The installer uses the length appropriate to the connector to be installed.

Remove this much jacket

▬▬▬▬▬▬▬ 1.25"

Leave this much buffer tube

▬▬▬ 0.625"

Leave this much aramid yarn

▬ 0.313"

Figure 6-4: Strip Lengths Template

Figure 6-5: Tubing Cutter On Cable

6.3.5.2.2 STRIP BUFFER TUBE

The installer holds the aramid yarns back against the remaining jacket. He marks the buffer tube length to be left (Figure 6-4). He strips the buffer tube and primary coating as in 6.3.5.1.

6.3.5.2.3 TRIM STRENGTH MEMBERS

The installer separates the aramid yarn from the buffer tube. He twists and loops the yarn over the scissor blade. He cuts the yarn to the length in Figure 6-4 (Figure 6-6).

Some connectors allow trimming the strength member as the second step. For such connectors, the strength member length must be less than the buffer tube length.

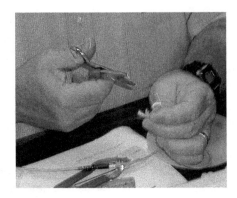

Figure 6-6: Cutting Aramid Yarn

6.3.5.3 CABLE CONTINUITY TEST

This test is optional, but recommended. This test works on both multimode and singlemode fibers to several thousand feet.

The installer holds a high intensity light source close to the fibers in one end of the cable. A 1 W single LED flashlight works well. An associate at the opposite end views the fibers. Each fiber core should glow, indicating continuity.

6.3.5.4 FIBER STRENGTH TEST

This step is optional. The installer performs this step to troubleshoot fiber breakage. The installer performs this step for each fiber just prior to inserting it into a connector.

The installer pushes each fiber against a lens grade tissue on the work surface. If the fiber bends without breaking, the installer has not damaged the cladding. If there is a single defect on the cladding, the fiber will break.

SAFETY WARNING: the installer should not push a bare fiber against his finger. Doing so can result in the fiber penetrating the skin and breaking.

6.3.6 CLEAN FIBER

Immediately prior to inserting each fiber into a connector, the installer performs this step. The installer moistens a lens grade tissue with isopropyl alcohol. He folds the moist area around the fiber (Figure 6-7). He pulls the fiber through the fold twice. He inspects the fiber for particles on the cladding. If necessary, he cleans the fiber repeatedly until the cladding is free of all particles.

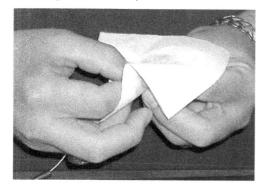

Figure 6-7: Cleaning Fiber With Moist Tissue

6.3.7   CONNECTOR TESTS

At this step, the installer can make two tests. Both these tests are troubleshooting steps for failure of fibers to fit into connectors or for fiber breakage. Both tests are optional. If necessary for troubleshooting or desirable for practice, the installer repeats these steps for all connectors.

6.3.7.1   WHITE LIGHT TEST

While aiming the ferrule towards a light source, such as a window or ceiling light, the installer looks into the back shell of a connector. He should see a sparkle that indicates the presence of a fiber hole and the absence of contamination in the connector. Such contamination could block the path of the fiber.

6.3.7.2   DRY FIT TEST

While resting his hands together, the installer feeds the fiber into central tube in the back shell of the connector (Figure 6-8). As soon as the fiber enters the back shell, he twists the connector back and forth. As long as the fiber continues feeding into the back shell without bending, he continues twisting the connector and inserting the fiber. If the fiber bends, he withdraws the fiber approximately 1/8", twists the connector back and forth and feeds the fiber into the back shell.

The installer must see the fiber protruding beyond the tip of the ferrule. This protrusion indicates that the fiber and the

hole are compatible, that the fiber is sufficiently long, and that there is nothing blocking the fiber path through the ferrule. The installer removes the fiber from the connector and places that connector near that fiber.

Note: if the installer places the fiber on a surface, he must repeat 6.3.6 prior to inserting the fiber into the connector.

Figure 6-8: Central Tube Of SC Connector

6.3.8   INJECT ADHESIVE

Before each use, the installer wipes the adhesive from the outside of the needle. He places the needle into the connector back shell until the needle butts against the inside end of the ferrule. While maintaining pressure on the needle, he presses the plunger until the adhesive flows through the fiber hole in the end of the ferrule. As soon as the adhesive flows through the fiber hole, he removes the needle from the connector. He pulls the plunger 1/8" out of the barrel of the syringe to eliminate 'weeping'. With experience, he can repeat this step for up to 6 connectors.

6.3.9   PRIMER APPLICATION

Using the brush from the primer bottle, the installer brushes the fiber and the last 2 mm of the buffer tube with the primer.

Caution
After wiping the fibers with primer, do not delay inserting the fiber. Excessive time allows the primer to harden, making the fiber too large to fit in the ferrule.

6.3.10 FIBER INSERTION

The installer inserts the fiber into the center tube in the back shell of a filled connector. As soon as the fiber enters the tube, he twists the connector back and forth. As long as the fiber continues into the tube without bending, he continues twisting the connector and inserting the fiber. If the fiber bends, the installer withdraws the fiber 1/8", twists the connector back and forth and feeds the fiber into the back shell.

When the buffer tube stops further motion of the fiber into the ferrule, he inspects the tip of the ferrule. He should see the fiber protruding beyond the tip of the ferrule. The amount of protrusion is not critical.

If he does not see the fiber protruding beyond the tip of the ferrule, he removes and discards the connector. If he cannot see the fiber protruding beyond the end of the ferrule, the fiber was not stripped to a sufficient length, the fiber broke during insertion, or the adhesive set prior to full insertion of the fiber.

The installer may tape the buffer tubes to the work surface so that the buffer tubes do not slide out of connectors. Should the cables slide out, the fibers may not protrude beyond the end of the ferrule or may break below the tip of the ferrule.

For premises or loose tube cables, the installer proceeds to 6.3.12.

6.3.11 CRIMP SLEEVE

While sliding the sleeve over the back shell, he rotates the sleeve back and forth. He checks the length of the strength members. They should be flush with the end of the crimp sleeve or less than 1/16" longer than the crimp sleeve.

He crimps the crimp sleeve with the appropriate crimp nest or nests (Figure 6-9). He crimps the large diameter of the sleeve over the back shell of the connector and the small diameter of the crimp sleeve over the jacket.

Caution: prior to crimping, check the position of the crimp sleeve: it should be in a nest, not between nests.

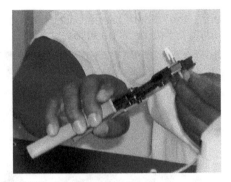

Figure 6-9: Crimping Crimp Sleeve

6.3.12 CREATE BEAD

This step is optional. The increase in bead size is minimal and may not provide sufficient increase in bead size to justify.

The installer dips the primer brush into the primer. While holding a connector with the fiber pointing up, he wipes the brush against the fiber. He allows primer to run down the fiber onto the bead of adhesive on the tip of the ferrule. With a lens grade tissue, he wicks excess primer from the tip of the ferrule. He does not touch or break the fiber.

6.3.13 REMOVE FIBER

While watching the fiber at the end of the ferrule, the installer pulls gently on the jacket or buffer tube. The fiber should not move. If it moves, the adhesive has not hardened. Once the adhesive has hardened, the installer proceeds.

While resting his hands together, the installer places the wedge surface of a scriber onto the ferrule end (Figure 6-10). He moves the blade of the scriber to the fiber so that the scriber gently touches the fiber at the top of the adhesive bead. The fiber should not move or bend.

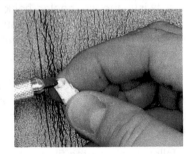

Figure 6-10: Scribing Excess Fiber

The installer moves the scriber along the fiber 1/16" once. He should not break with the scriber or 'saw' the fiber with the scriber.

The installer holds the connector with the fiber pointing up. He grips the connector with his thumb and forefinger lightly. He slides his thumb and forefinger up the connector towards and over the fiber. He grips and pulls the fiber away from the tip of the connector. The fiber should break easily. He places of the fiber in the fiber collection bottle.

> Note: the installer does not try to grasp the fiber from the side. Doing so can break the fiber in bending, not in tension. If the fiber breaks in bending, it may break below the surface of the connector. With ceramic ferrules and standard polishing films, polishing does not remove this condition.

If the fiber does not break, the installer scribed adhesive on the fiber and not the fiber. He repeats this step until the fiber breaks. The installer wipes the scribing blade to remove any uncured adhesive.

With a lens grade tissue, the installer wipes the scriber blade to remove any primer or uncured adhesive.

6.3.14 AIR POLISH

The installer holds the connector with the ferrule pointing up. He holds a 12-µ polishing film at one edge with the dull (abrasive) side down. He places the opposite edge of the film above the connector. He curls the front and back edges of the film down (Figure 6-11). With a light pressure, he rubs the film against connector until the fiber is flush with the bead of adhesive. If he scribed the fiber just above the surface of the adhesive bead, this step will take less than 10 seconds. The installer does not remove all the adhesive.

The installer checks the fiber by bringing a finger down onto the tip of the ferrule from the top. If he cannot feel the fiber, the fiber is flush, or nearly flush, with the adhesive.

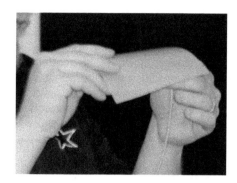

Figure 6-11: Air Polishing Fiber

> Caution: do not rub a finger across the tip of the ferrule. If the fiber is not flush with the adhesive, the installer may snag and break the fiber. In addition, the broken fiber may pierce the installer's skin.

6.3.15 MULTIMODE POLISH

For single connectors, the installer follows the procedure in this section for that connector. For multiple connectors, the installer can perform each step on all connectors prior to performing the next step; i.e., a 'batch process'.

6.3.15.1 CLEAN CONNECTORS

The installer moistens a lens grade tissue with isopropyl alcohol. He wipes the ferrules of all connectors with this moistened tissue. This removes 12-µ grit.

6.3.15.2 CLEAN EQUIPMENT

With lens-grade compressed gas or isopropyl alcohol, he cleans off the top surface of the polishing pad(s), both sides of the polishing film(s), and the polishing tool. He places the polishing film(s), dull side up, on the pad(s).

6.3.15.3 FIRST POLISH

The purpose of the first polish is removal of the adhesive. Additional time after such removal will not improve the loss, and can result in increased loss.

The installer places the polishing tool, hereafter 'tool', on the 3-µ film. He inserts the connector into the polishing tool.

Light pressure means that the installer holds the connector against the film without compressing the spring inside the SC connector.

While holding both the connector and the tool (Figure 6-12) with a light pressure, he moves the tool in a ½" high, figure-8 pattern slowly. If he feels scratchiness, the fiber is not flush with the adhesive. He maintains a light pressure and a slow, ½" high figure-8 pattern until the scratchiness ceases.

Figure 6-12: Polishing

After the scratchiness ceases, the installer increases the size of the figure-8 pattern to cover the entire film. He does not increase the pressure. He continues polishing until the adhesive is completely removed.

The installer may detect a change in the friction of the connector on the film. This change indicates the removal of the last of the adhesive. If he does not detect this change, he removes the connector from the tool periodically. He views the tip of the ferrule by reflecting light off the tip. When the adhesive is completely removed, the tip of the ferrule will be glass smooth and shiny. The tip will have the same appearance as that of the side of the ferrule.

6.3.15.4   CLEANING

With lens grade compressed gas or isopropyl alcohol and lens grade tissues, the installer cleans the connector and the polishing tool.

6.3.15.5   SECOND POLISH

The purpose of the second polish is creation of a 'featureless,' low loss core (4.4). 'High pressure' for polishing means holding the connector in the polishing tool and pushing the inner housing down until the internal spring is fully compressed.

The installer need not increase the pressure beyond the level at which the spring is fully compressed.

The installer places the polishing tool on the 1-μ film and applies high pressure to the connector for ten, large figure-8 patterns. Large means using the full area of the film.

6.3.15.6   THIRD POLISH

This polish is optional. Polish as in 5.3.16.6.

6.3.16 SINGLEMODE POLISHING

Use the procedure in 5.3.21.

6.4   FINAL CLEANING

He cleans the connector with one of the following methods.

Do not use lens grade gas for final cleaning of the connector. This gas can leave watermarks that increase loss and reflectance.

6.4.1   BEST METHOD

The installer moistens a lens grade, lint free tissue with isopropyl alcohol. He wipes the sides and tip of the ferrule with this tissue. He sprays a one-inch diameter area of connector cleaner, such as ElectroWash Px, onto a small pad of lens grade tissues. Three times, he wipes the tip of the ferrule from the wet area to the dry area.

Optional: with lens grade gas, the installer blows out the connector cap.

He installs the cap.

6.4.2   METHOD B

The installer moistens a tissue with isopropyl alcohol. He wipes the sides and the tip of the ferrule. Immediately, he wipes the tip of the ferrule with a dry tissue.

Optional: with lens grade gas, the installer blows out the connector cap.

He installs the cap.

6.4.3 METHOD C

The installer wipes the sides and the tip of the ferrule with a pre-moistened lens grade tissue (Alco Pad or OptiPrep Pad). Immediately, he wipes the tip of the ferrule with a dry tissue.

Optional: with lens grade gas, the installer blows out the connector cap.

He installs the cap.

6.4.4 INSPECTION

After polishing each connector, the installer inspects and rates the connector as in 4.4. In addition, he inspects the cladding for uniformity of appearance (Figure 4-6).

If the connector is not perfect, he tags the connector. If the connector is 'terrible' (4.4.5), he replaces the connector.

Note: this author does not recommend polishing more than one connector prior to microscopic inspection. Should the film be contaminated, cores can become scratched. Discovering such contamination after one connector is better than after a dozen!

6.4.5 WHITE LIGHT TEST

When the installer has installed all connectors on the one end of a cable, he performs a white light, continuity test as in 6.3.7.1 (Figure 6-13). The installer does not install connectors on the opposite end until all of the fibers have passed this continuity test. If one or more connectors fail this test, he troubleshoots these connectors to identify the problem. An OTDR test may be required.

The installer replaces the caps on all connectors that pass this test. He repeats this continuity test after he installs all connectors on the opposite end of the cable.

6.5 FINAL ASSEMBLY

The installer slides the boot over the back shell of the connector. He tugs on the jacket or the buffer tube gently.

The top of the inner housing has missing corners (Figure 6-14). The top of the outer

housing has the key. He holds the top of the inner housing and the top of the outer housing up. He inserts the inner housing into the end of the outer housing opposite the key (Figure 6-15). He wiggles the inner housing until it snaps through the outer housing (Figure 6-16).

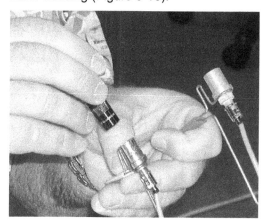

Figure 6-13: White Light Test

Figure 6-14: Alignment Of Inner and Outer Housings For Final Assembly

Figure 6-15: Insertion Of Inner Housing

Figure 6-16: Inner Housing Fully Inserted Into Outer Housing

6.6   TROUBLESHOOTING

6.6.1   INSTALLATION

6.6.1.1   FIBER BREAKAGE

Multimode fiber breaks upon insertion or does not fit into connector

Potential cause: dirt or primary coating on fiber
Action: clean fiber

Potential cause: debris in connector
Action: replace connector

If the connector fails this test, he blows out the connector with lens grade gas. If the connector still fails this test, he flushes the connector with isopropyl alcohol injected through connector with a syringe. If connector fails after this flush, the installer replaces the connector.

Potential cause: primary coating not completely removed
Action: re-strip the fiber

Potential cause: primer has dried
Action: do not delay inserting fiber into connector after wiping primer on fiber; prepare new end

Potential cause: adhesive has partially hardened
Action: reduce number of filled connectors

6.6.1.2   ADHESIVE FLOWS FROM BACK SHELL

Potential cause: excessive adhesive in back shell
Action: after adhesive flows from fiber hole in ferrule, immediately withdraw needle from connector

6.6.1.3   ADHESIVE ON OUTSIDE OF BACK SHELL

Potential cause: failure to wipe needle before each use
Action: wipe adhesive from outside of needle before each use

6.6.2   POLISHING

6.6.2.1   NON-ROUND CORE AND ABRASIVE TORN FROM FILM BACKING

Potential cause: incomplete air polishing
Action: increase air polish time

Potential cause: excessive pressure during polishing first film
Action: reduce polishing pressure

Potential cause: small bead of adhesive sheared from connector on first film
Action: replace connector; apply primer to end of ferrule

Potential cause: excessive air polishing; all adhesive removed; ferrule tip is mirror smooth without any dull or colored film or bead of adhesive
Action: reduce air polish time

6.6.2.2   EXCESSIVE POLISHING TIME ON FINAL FILM

Potential cause: film worn out
Action: replace film

6.6.2.3   EXCESSIVE POLISHING TIME

Potential cause: insufficient polishing pressure
Action: increase polishing pressure

Potential cause: film worn out
Action: replace film

6.6.2.4   NO APPARENT REDUCTION OF BEAD SIZE

Potential cause: dirt in ferrule hole of polishing tool
Action: clean hole in polishing tool with lens grade air or with pipe cleaner dipped in isopropyl alcohol

6.6.2.5   FEW CORE AND CLADDING SCRATCHES

Potential cause: contamination of film by environment
Action: replace film; move polishing location away from air vents or any other source of airborne dust

Potential cause: incomplete polishing
Action: complete polishing

6.6.2.6   ADHESIVE SMEARS
Potential cause: adhesive past its expiration date
Action: replace adhesive

6.6.2.7   CRACKED FIBER
Potential cause: excessive pressure during scribing
Action: scratch fiber lightly during scribing

6.6.3   SINGLEMODE POLISHING
See 5.4.3.

6.7   SUMMARY
This summary is in three parts, one each for connector installation, multimode polishing and finish installation.

6.7.1   INSTALLATION
Install cable through enclosure

Pull cable through enclosure to work surface

Remove jacket

Install boots

Install crimp sleeves on jacketed fibers

Trim strength members to proper lengths

Strip buffer tube

Perform cable continuity test

Transfer adhesive to syringe

Clean the fiber

Optional step: perform a white light test

Optional step: perform a dry fit

Inject adhesive

Wipe primer onto fiber and end of buffer tube

Twist and insert fiber into connector

Slide crimp sleeve over back shell

Crimp the crimp sleeve

Wipe fiber with primer

Scribe fiber

Remove excess fiber

Air polish excess fiber flush with adhesive bead

6.7.2   MULTIMODE POLISH
Clean connector

Clean pads, films and polishing tool

Polish on 3-µ film

Clean connector and polishing tool

Polish on 1-µ film

Clean connector and polishing tool

Optional: polish on 0.5-µ film

6.7.3   FINISH INSTALLATION
Clean connector

Inspect connector

White light cable test

Install outer housing

Clean and install cap

7 HOT MELT ADHESIVE INSTALLATION

Chapter Objectives: you install multimode, Hot Melt adhesive, ST-compatible connectors with low loss and high reliability on 900-µ tight tube cables.

7.1 INTRODUCTION

This chapter applies to the installation of ST™-compatible, connectors onto fibers in 900-µ tight tubes. This tube can be in a premises cable or a loose tube cable with furcation tubes on each fiber.

Installation of connectors by this method on other cable types requires end preparation appropriate to the cable design.

In addition, this chapter contains certain instructions for installation of the multimode SC connector. The SC connector method differs in three aspects:

> Design of the holder

> Method of loading connector into holder

> Installation and crimping of crimp sleeve on premises cable

This method of connector installation is a hot melt adhesive. This adhesive is preloaded into the connector. The connector requires pre-heating prior to and cooling after installation.

This method reduces installation time. This method allows for salvage of damaged connectors through reheating. In addition, the preloaded adhesive creates a bead of adhesive on the tip of the ferrule. This bead has a consistent, predetermined size. These latter two characteristics enable the novice installer to achieve high process yield. Typical yield by first-time installers is 85-90 %. In field installations, reheating can result in a 100% yield.

7.2 MATERIALS

The tools and supplies listed herein are required for installation of this connector by this method. We have chosen some of these tools and supplies on the basis of their superior performance. We have chosen others, on the basis of their price. Finally, we have chosen some of these

tools and supplies based on their convenience.

Sources of equipment and supplies are: Fiber Optic Center Inc. (FOCI) 800-ISFIBER and Fiber Instrument Sales (FIS), 800-5000-FIS.

For installation of these connectors, the installer requires a connector installation kit, which includes the following:

> 3M Hot Melt, ST™-compatible multimode connectors, part number 6100

> Alternate connector: multimode SC, part number 6300W

> Crimper for SC connector, with 0.190" and0.137" nests

> 400 x connector inspection microscope with ST-compatible or SC fixture (Westover Scientific)

> Work mat (Clauss Fiber-Safe™)

> Tubing cutter (Ideal 45-162)

> Kevlar scissors (with ceramic blade)

> Miller buffer tube and primary coating stripper (FO-103-S) or Clauss NoNik stripper (NN203)

> Hot Melt cooling stand

> Hot Melt holders for ST™-compatible (or SC) connectors

> Hot Melt oven

> 2-µ polishing film for Hot Melt connectors (FIS, 3M part number 51144 85932, 254X Imperial, 6192A)

> Lens grade tissue (Kim-Wipes™ or equivalent)

> 99% isopropyl alcohol (FIS)

> Alternative to isopropyl alcohol: Alco pads or Opti-Prep Pads

---

> ➤ Connector cleaner (Electro Wash® Px, from ITC Chemtronics)

> ➤ Wedge scriber (Corning Cable Systems Ruby Scribe, 3233304-01)

> ➤ Small plastic bottle (for bare fiber collection)

> ➤ Lens grade compressed gas (Stoner part number 94203 or equivalent)

> ➤ ST™-compatible polishing tool (FIS or FOCI)

> ➤ Two hard rubber polishing pads, (FIS part number PP 575)

> ➤ 12-µ polishing film (FOCI part number AO12F913N100)

> ➤ For multimode polishing: 1.0-µ polishing film (FOCI part numbers AO1T913N100)

Scissors with ceramic blades provide the best results, but have the highest cost. Scissors with other types of blades will work. The polishing tool is also called a fixture and a puck. No pad is required for the 12-µ air polish film.

Some hot melt ovens have a cover that functions as a cooling stand. The films listed are for multimode polishing.

7.3   PROCEDURE

7.3.1   PREINSTALL CABLE

The installer installs the cable into the enclosure without permanently attaching the cable to the enclosure. He pulls the cable out of the enclosure to a work surface near the enclosure. After installation of all connectors, he completes the installation of the cable in the enclosure.

7.3.2   SET UP OVEN

The installer turns on the preheating oven and allows it to heat up for at least 15 minutes. He assembles the cooling stand (Figure 7-1). Note that current generation 3M cooling stands require no assembly.

Note that some hot melt ovens have the cooling stand function built into the top surface of the oven. Such ovens can be used, by shutting the oven off, when top surface becomes too hot to allow cooling in a short time.

Figure 7-1: Hot Melt Cooling Stand

7.3.3   LOAD HOLDERS

7.3.3.1   GENERATION 1 HOLDERS

The installer removes the connector parts from the packages. He removes the connector caps. He aligns the key of a connector with the slot in a connector holder. He slides the connector into the holder. He rotates the retaining nut so that the holder retains the connector (Figure 7-2). The installer loads 4-6 connectors into holders. He places the holders into the cooling stand.

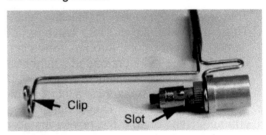

Figure 7-2: Loaded Connector Holder

To load SC holders, the installer aligns the two flat surfaces of the SC inner housing with the fingers of the holder, inserts the housing into the holder until the spring is fully compressed, and rotates the housing 90°. The holder grips the connector [Figure 7-3].

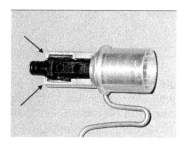

Figure 7-3: Loaded SC Hot Melt Holder

7.3.3.2   CURRENT HOLDERS

Current generation holders do not require alignment of key of the ST™-compatible connector or rotation for the SC connector.

7.3.4   REMOVE OUTER JACKET

The installer removes the outer jacket of the premises cable by the appropriate procedure.

7.3.5   INSTALL BOOTS

7.3.5.1   PREMISES CABLE

The installer installs the boot small end first, and a clear flexible tube on each buffer tube. He pushes the boots and tubes approximately 12" away from end of buffer tubes. He marks the length of buffer tube to be removed (Figure 7-4 and Figure 7-5).

▬▬▬▬▬

Novice installers, 0.87"

▬▬▬▬▬

Experienced installers, 1.06"
Strip this much bare fiber

Figure 7-4: ST™-compatible Template
For Premises Cable (to scale)

▬▬▬

Novice installers, 0.56"

▬▬▬▬

Experienced installers, 0.81"
Strip this much bare fiber

Figure 7-5: Template For SC Connector
And Premises Cable

Note: the 3M instructions for the ST™-compatible connector indicate the length

of 0.87". However, installers experienced with polishing can reduce polishing time by using the increased strip length. This increased length removes some of the hot melt adhesive from the ferrule tip, thus reducing polishing time. However, this author does not recommend stripping to a length greater than 1.5", as such a length can result in fiber breakage during insertion.

7.3.6   PREPARE END

The installer repeats this step for all fibers to be terminated at this location.

7.3.6.1   NO-NICK STRIPPER

If the installer uses a No-Nick® Stripper, he cleans the stripper by holding the stripper open and snapping each cap once (Figure 7-6).

Occasionally, this method of cleaning does not work. In that case, the installer blows out the stripper with lens grade gas.

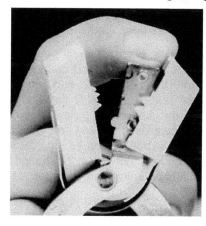

Figure 7-6: Cleaning Clauss Stripper

The installer grips the cable or buffer tube firmly. The installer can grip the cable by wrapping the cable around one finger 5 times (Figure 7-7) or by weaving the cable through his fingers. The installer places the stripper on the buffer tube at the marked distance (Figure 7-7) or 1/2" from the end of the buffer tube, whichever distance is less. While holding the stripper at 90° to the fiber (Figure 7-7), the installer pulls the stripper *in the direction of the arrow* on the stripper towards the fiber end slowly. He *allows* the buffer tube and primary coating to slide from the fiber.

He does not force the buffer tube from the fiber.

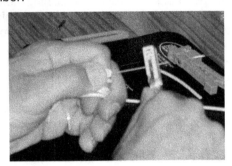

Figure 7-7: Fiber Straight In Stripper

This instruction indicates a maximum strip length of 1/2". US-made fiber optic cables can be stripped to at least this distance. Some cables allow a longer strip length without breakage. For such cables, you may strip to that increased distance. Few cables allow more than 1.25 ".

7.3.6.2   MILLER STRIPPER

He cleans the Miller stripper with lens grade gas. The installer grips the buffer tube firmly as described in 7.3.6.1.

The installer places the stripper on the buffer tube at the marked distance or at 1/2" from the buffer tube end, whichever distance is less. He holds the stripper at 45° to the fiber. He pulls the stripper slowly towards the end of the fiber, *allowing* the buffer tube to slide from the fiber. He does not force the buffer tube and primary coating from the fiber.

If necessary, the installer repeats this step until the length of bare fiber is as indicated in Figure 7-4 or Figure 7-5 (for premises cable) or the length of bare fiber appropriate for jacketed cable. He inspects the fiber to ensure that there is no buffer tube or primary coating remaining on the fiber.

Note that primary coating may extend beyond the end of the buffer tube. If the length of the primary coating is less than 3/16", the installer can leave it. The hot melt adhesive will melt the primary coating, allowing full insertion of the fiber.

7.3.6.3   CABLE CONTINUITY TEST

This test is optional, but recommended. This test works on both multimode and singlemode fibers to several thousand feet.

The installer holds a high intensity light source close to the fibers in one end of the cable. A 1 W single, LED flashlight works well. An associate at the opposite end views the fibers. Each fiber core should glow, indicating continuity.

7.3.6.4   FIBER STRENGTH TEST

This step is optional to troubleshoot breakage. The installer performs this step for each fiber just prior to inserting it into a connector.

The installer pushes each fiber against a lens grade tissue on the work surface. If the fiber bends without breaking, the cladding is defect free. If there is a single defect on the cladding, the fiber will break.

> SAFETY WARNING: the installer should not push a bare fiber against his finger. Doing so can result in the fiber penetrating the skin and breaking.

7.3.7   CLEAN FIBER

The installer performs this step for each fiber immediately prior to inserting it into a connector. The installer moistens a lens grade tissue with isopropyl alcohol. He folds the moist area around the fiber (Figure 7-8). He pulls the fiber through the fold twice. He inspects the fiber for particles on the cladding. If necessary, he cleans the fiber repeatedly until the cladding is free of all particles.

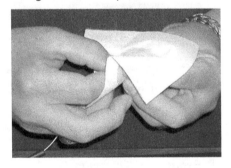

Figure 7-8: Cleaning Fiber

7.3.8   PRE-HEATING

The installer places the loaded connector holders into the oven so that the wide flange of the holder rests against the oven-heating block (Figure 7-9). If the hot melt adhesive bubbles out of the back shell, he discards the connector: the connector has absorbed excessive moisture. After allowing the connector to heat for the required time, he removes one holder from the oven. If the adhesive is blue, the pre-heat time is 60 seconds; if red, 90 seconds.

> Caution: the holders are hot enough to burn fingers severely!

7.3.9 INSERT FIBER

He verifies that the boot and clear tubing are still on the jacket or buffer tube. The installer holds the connector holder so that he can insert the fiber into the back shell. He holds the cable near the end of the jacket or the end of the buffer tube. While twisting the cable or buffer tube back and forth slowly, he inserts the end of the fiber into the back shell. As long as the fiber continues feeding into the back shell without bending, he continues twisting and inserting the fiber.

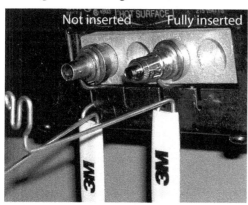

Figure 7-9: Connectors In Oven

For a jacketed fiber on an ST™-compatible connector, the jacket will enter the back shell. The aramid yarn will fold over the jacket and enter the back shell. A small drop of hot melt adhesive may come out of the back shell. This is normal.

If the fiber bends, the installer withdraws the fiber 1/8", twists the fiber back and forth and inserts the fiber into the back shell. When the buffer tube stops further motion of the fiber into the ferrule (Figure 7-10), he presses the cable into a 'V' area

of the holding clip (Figure 7-11). Note that the SC holder does not have a clip.

For a premises cable, the installer slides the clear tubing into the back shell and the boot over the tubing and back shell. Without putting tension on the buffer tube, he places the holder in the cooling stand (Figure 7-1).

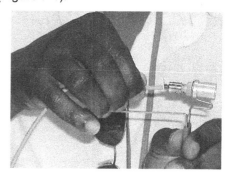

Figure 7-10: Cable Inserted Into Connector

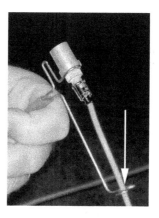

Figure 7-11: Cable In Clip

7.3.10 REMOVE EXCESS FIBER

The installer touches the holder carefully to ensure that it is cool enough to handle. The method of removal is the reverse of that for loading. Without breaking the fiber protruding beyond the bead of adhesive, the installer removes the connector from the holder. He should see the fiber protruding beyond the bead of adhesive.

While resting his hands together, the installer places the wedge surface of a scriber onto the ferrule end (Figure 7-12). He moves the blade of the scriber to the fiber so that the scriber gently touches the fiber at the top of the adhesive bead. The fiber should not move or bend.

The installer moves the scriber back and forth along the fiber 1/16" once. He should not break the fiber with the scriber. He should not 'saw' the fiber with the scriber.

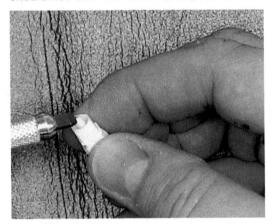

Figure 7-12: Scribing Excess Fiber[11]

The installer holds the connector with the fiber pointing up. He grips the ferrule with his thumb and forefinger lightly. He slides his thumb and forefinger up the ferrule towards and over the fiber. He grips and pulls the fiber away from the tip of the ferrule. The fiber should break easily. He places of the fiber in the fiber collection bottle.

If the fiber does not break, the installer scribed the hot melt adhesive on the fiber and not the fiber. He repeats this step until the fiber breaks.

7.3.11 AIR POLISHING

The installer holds the connector with the ferrule pointing up. He holds a 12-µ polishing film at one edge with the dull (abrasive) side down. He places the opposite edge of the film above the connector. He curls the front and back edges of the film down (Figure 7-13). With a light pressure, he rubs the film against connector until the fiber is flush with the bead of adhesive. If he scribed the fiber just above the surface of the adhesive bead, this step will take less than 10 seconds. The installer does not remove all the adhesive.

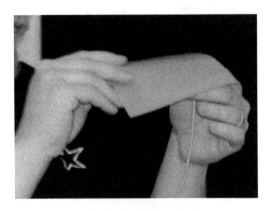

Figure 7-13: Air Polishing Fiber

The installer checks the fiber by bringing a finger down onto the tip of the ferrule from the top. If he cannot feel the fiber, the fiber is flush, or nearly flush, with the adhesive.

Caution: do not rub a finger across the tip of the ferrule. If the fiber is not flush with the adhesive, the installer may snag and break the fiber. In addition, the broken fiber may pierce the installer's skin.

7.3.12 MULTIMODE POLISHING

For single connectors, the installer follows the procedure in this section for that connector. For multiple connectors, the installer can perform each step on all connectors; i.e., a 'batch process'.

7.3.12.1   CLEAN CONNECTORS

The installer moistens a lens grade tissue with isopropyl alcohol. He wipes the ferrules of all connectors with this moistened tissue. This removes 12-µ grit.

7.3.12.2   CLEAN EQUIPMENT

With lens-grade com-pressed gas or isopropyl alcohol, he cleans off the top surface of the polishing pad(s), both sides of the polishing film(s) and the bottom of the polishing tool. He places the polishing film(s) on the pad(s), dull side up.

7.3.12.3   FIRST POLISH

The purpose of the first polish is removal of the adhesive. Additional time after such removal will not improve the loss and can result in increased loss.

---

[11] We have shown the SC connector.

The installer places the tool on the 2-μ film. The installer inserts the connector into the polishing tool. While holding both the connector and the tool (Figure 7-14), he moves the tool in a ½" high, figure-8 pattern slowly. If he feels scratchiness, the fiber is not flush with the adhesive. He maintains a light pressure and a slow, ½" high figure 8 pattern until the scratchiness ceases.

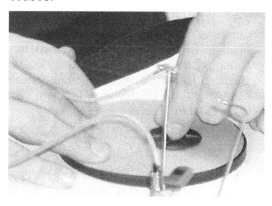

Figure 7-14: Polishing

After the scratchiness stops, the installer increases the size of the figure-8 pattern to cover the entire area of the film. He *does not increase* the pressure. He can increase the speed. He continues polishing until the adhesive is completely removed.

The installer may detect a change in the friction of the connector on the film. This change indicates the removal of the last of the adhesive.

If he does not detect this change, he removes the connector from the tool periodically. He views the tip of the ferrule by reflecting light off the tip. When the adhesive is completely removed, the tip of the ferrule will be glass smooth and shiny. The tip will have the same appearance as that of the side of the ferrule.

With experience, the installer recognizes the normal size of the fiber hole. If the hole appears larger than it is normally, there is a small amount of adhesive remaining on the connector (4.6, Review Questions 15 and 23). A few additional polishing strokes removes this adhesive.

7.3.12.4 CONNECTOR CLEANING

With lens grade compressed gas or isopropyl alcohol and lens grade tissues, the installer cleans the connector, and the polishing tool.

7.3.12.5 SECOND POLISH

This polish improves microscopic appearance and simplifies inspection (4). This polish will not reduce multimode loss.

'High pressure' for polishing means holding the ST™-compatible connector by its retaining nut and pushing the retaining nut down until the internal spring is fully compressed. The installer need not increase the pressure beyond this level. For an SC connector, the installer pushes the inner housing down until the spring is fully compressed.

The installer applies high pressure to the connector for ten, large figure 8 patterns. Large means using the full area of the polishing film.

7.3.13 SINGLEMODE POLISHING

See manufacturer instructions for this step. Singlemode polishing requires replacement of the 1-μ, second polishing film, with a 0.05-μ film. Final polishing requires three figure eights with a wet polish.

7.4 SC CRIMP SLEEVE

For installation of the SC connector, the installer slides the crimp sleeve over the back shell. He slides the clear plastic tubing into the crimp sleeve. He crimps both the large and small diameters of the crimp sleeve with the 0.190" and 0.137" nests, respectively.

7.5 FINAL CLEANING

He cleans the connector with one of the following methods. Do not use lens grade gas for final cleaning of the connector. This gas leaves watermarks that can increase loss and reflectance.

7.5.1 BEST METHOD

The installer moistens a lens grade, lint free tissue with isopropyl alcohol. He wipes the sides and tip of the ferrule with this tissue. He sprays a one-inch diameter area of connector cleaner, such as ElectroWash Px, onto a small pad of lens grade tissues. Three times, he wipes the tip of the ferrule from the wet area to the dry area.

Optional: with lens grade gas, the installer blows out the connector cap.

He installs the cap

7.5.1.1 METHOD B

The installer moistens a tissue with isopropyl. He wipes the sides and the tip of the ferrule. Immediately, he wipes the tip of the ferrule with a dry tissue.

Optional: with lens grade gas, the installer blows out the connector cap.

He installs the cap.

7.5.1.2 METHOD C

The installer wipes the sides and the tip of the ferrule with a pre-moistened lens grade tissue (Alco Pad or OptiPrep Pad). Immediately, he wipes the tip of the ferrule with a dry tissue.

Optional: with lens grade gas, the installer blows out the connector cap.

He installs the cap.

7.6 CAP INSTALLATION

Optional: the installer blows out the cap with lens grade gas.

The installer installs the cap.

7.7 INSPECTION

After polishing each connector, the installer inspects and rates the connector as in 4.4. In addition, he inspects the cladding for uniformity of appearance (Figure 4-6).

If the connector is not perfect, he tags the connector. If the connector is 'terrible' (4.4.5), he replaces the connector.

Note: this author does not recommend polishing more than one connector prior to microscopic inspection. Should contamination be on the film, cores can become scratched. Discovering such contamination after one connector is better than after a dozen!

7.8 WHITE LIGHT TEST

When the installer has installed all connectors on the one end of a cable, he performs a white light, continuity test as in 7.3.6.3 (Figure 7-15). The installer does not install connectors on the opposite end until all of the fibers have passed this continuity test. If one or more connectors fail this test, he troubleshoots these connectors to identify the problem. An OTDR test may be required.

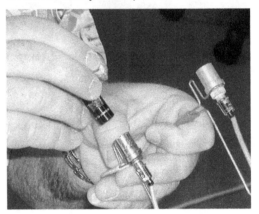

Figure 7-15: White Light Test

The installer replaces the caps on all connectors that pass this test. He repeats this continuity test after he installs all connectors on the opposite end of the cable.

7.9 SC FINAL ASSEMBLY

See 5.3.20 for final assembly.

7.10 SALVAGE

7.10.1 PROCEDURE

When the installer installs the hot melt, ST™-compatible connector onto premises cables or the 3 mm jacketed cable, he can reheat the connector to repair damage. This salvage can be performed as long as the installer has not performed either of the following actions:

> ➤ Crimped the crimp sleeve (SC only)

> ➤ Installed the SC outer housing over the inner housing (SC only)

To salvage a damaged fiber end, the installer places the connector into the holder, reheats the connector, removes the cable, prepares the cable end, and inserts the new end. A new end is ready for polishing.

When the installer salvages a Hot Melt connector, the new end has an adhesive bead smaller than during the first installation. This small bead provides reduced support to the fiber during polishing. Because of this bead size reduction, this salvage procedure requires a change to the scribing and polishing techniques.

When scribing, the installer holds the scriber so that the blade touches the fiber slightly above the surface of the bead. After scribing, he air polishes without removing all the adhesive. He polishes with a modified technique: he uses a very light pressure, a ½" high very slow, 'figure 8' movement until the bead is completely removed. He polishes as in 7.3.12.5.

7.10.2 MODIFIED PROCEDURE

The installer can eliminate the re-preparation for salvage by modifying the procedure in 7.3.9. Instead of fully inserting the fiber, he inserts the fiber fully, and then withdraws the cable by approximately 1/16". This modification leaves extra bare fiber buried in the adhesive. After reheating, he pushes this extra fiber through the ferrule to create a new end for polishing. However, the adhesive bead will be very small.

This small bead size has two consequences. First, the installer performs air polishing without removing all of the adhesive bead. Second, the installer polishes on the 2-µ film very slowly to avoid snagging and breaking the fiber.

7.11 TROUBLESHOOTING

7.11.1 INSTALLATION

7.11.1.1 FIBER BREAKAGE
Multimode fiber breaks upon insertion into connector
Potential cause: dirt on fiber
Action: clean fiber

Fiber does not fit into connector
Potential cause: primary coating not completely removed
Action: restrip the fiber

Potential cause: adhesive is too cool for insertion.
Action: reheat connector

7.11.1.2 ADHESIVE FLOWS FROM BACK SHELL

Potential cause: moisture in adhesive
Action: discard connector

7.11.2 POLISHING

7.11.2.1 NON-ROUND CORE
Potential cause: broken fiber due to excessive polishing pressure
Action: check the 2-µ polishing film. If abrasive has been torn from backing, fiber protruding above adhesive snagged on film and broke. Increase air polish time

Potential cause: excessive pressure during polishing sheared off bead of adhesive. Check 2-µ film for small bead of adhesive
Action: reduce polishing pressure

Potential cause: all adhesive removed by excessive air polishing
Action: check the connector before 2µ polishing. If the tip of the ferrule is mirror smooth without any dull or colored film or bead of adhesive, no adhesive remains
Action: salvage connector and reduce air polish time

7.11.2.2 EXCESSIVE POLISHING TIME ON EITHER FILM

Potential cause: film worn out
Action: replace film

7.11.2.3 NO APPARENT REDUCTION OF BEAD SIZE

Potential cause: dirt in ferrule hole of polishing tool prevents tip of ferrule from contacting film

Action: clean hole in polishing tool with lens grade air or with pipe cleaner dipped in isopropyl alcohol

7.11.2.4 A FEW CORE AND CLADDING SCRATCHES

Potential cause: contamination of film by environment

Action: replace film. Clean film only if no replacement film is available.

Action: move polishing location away from air vents or any other source of airborne dust

7.11.2.5 A FEW SCRATCHES ON CORE

Scratches extend across core and cladding

Potential cause: contamination of film
Action: replace film. Clean film only if no replacement film is available.

Potential cause: incomplete polishing
Action: complete polishing

7.11.2.6 CRACKED FIBER

Potential cause: excessive pressure during scribing
Action: scratch fiber lightly during scribing

7.11.2.7 HIGH LOSS

With no cause apparent from microscopic inspection

Potential cause: over polished on either film.
Action: polish of 2-µ film only until adhesive is gone. Polish on the 1-µ film for 10 strokes

7.12 ONE PAGE SUMMARY

This summary is in two parts, one each for connector installation and multimode polishing

7.12.1 INSTALLATION

Install cable through enclosure

Pull cable through enclosure to work surface

Set up oven and cooling stand

Load connectors into holders

Remove jacket, Kevlar and central strength member to proper lengths

Perform cable continuity test

For ST-™ compatible and SC connectors and 900-µ tight tube, install boots and clear plastic tube on all buffer tubes. For jacketed cables, install boots and crimp sleeves

Strip buffer tubes to proper lengths

Preheat connector

Clean the fiber with lens grade tissue and isopropyl alcohol

With rotation, insert fiber into back shell until fiber bottoms out against ferrule

Do not allow fiber to bend

For jacketed cable and SC connectors, slide the crimp sleeve or boot over back shell

For jacketed fibers, press the jacket into the clip of the connector holder

For jacketed fiber. crimp the crimp sleeve

Scribe and remove the excess fiber

Air polish the excess fiber flush with the adhesive

Proceed to multimode or singlemode polishing procedure

7.12.2 MULTIMODE POLISHING

Clean all connectors

Clean the connector ferrule, pad, film and tool.

Place 2-µ film on the pad and the tool on the film

Install the connector into the tool

Slowly polish in a ½" high figure 8 pattern with light pressure until the scratchiness ceases

Polish in large figure 8 pattern with light pressure until all the adhesive is gone, using the entire area of the film

Clean the 1-µ film, the tool and the connector

Install the connector into the tool With high pressure, make 10 large figure 8 motions

Clean the side and tip of the connector

Clean and install the cap

Inspect the connector

Perform a white light test of the cable

After installing connectors on the second end, perform a white light test of the cable

8   CLEAVE AND CRIMP INSTALLATION

Chapter Objectives: you install SC and ST-™ compatible connectors by a 'cleave and crimp' method and achieve low loss and low cost.

8.1   INTRODUCTION

These Corning Inc. products, the UNICAM™ connectors, require no adhesive and no polishing. This installation method has the lowest installation time of all methods. While this method is fast, its' apparent simplicity hides subtleties. These subtleties determine the insertion loss, the process yield, and the total installed cost. Attention to these subtleties results in consistent and acceptable loss and yield.

At the time of this writing, Corning has developed at least four generations of installation tools. The fourth generation consists of a VFL-crimper tool (Figure 8-1), an improved cleaver (Figure 8-2) or the latest generation cleaver (Figure 8-3).

The VFL tool injects light into the connector. When a fiber in inserted into the connector, this tool detects any light that escapes from the connector. With colored LEDs, the tool indicates that the escaping light power level is acceptable or excessive. This author has worked with this tool for the last three years. It works well enough that no additional instructions are needed.

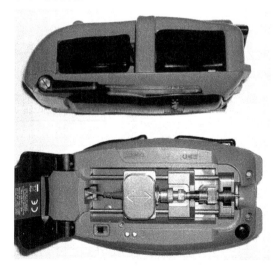

Figure 8-1: Unicam™ VFL-Crimper Tool

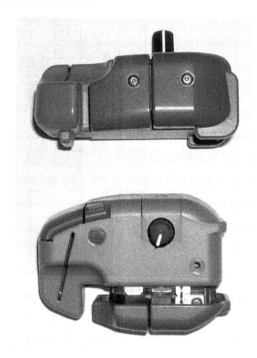

Figure 8-2: Unicam™ Cleaver

Figure 8 3: Latest Unicam Cleaver

In this chapter, we provide the instructions for use of earlier generation tools with a high quality cleaver.

The procedure herein is based on a high quality cleaver, the Alcoa Fujikura CT04 (Figure 8-4). This cleaver is different from the two cleavers recommended by the manufacturer. The CT07 produces low cleave angles, low loss connectors, high yield, low total installed cost and high consistency in loss.

Figure 8-4: CT04 Cleaver

8.2 MATERIALS

➢ Jacket stripper, Ideal tool (45-162)

➢ Tight tube and primary coating stripper (Clauss No-Nick® 203-μ or Miller stripper)

➢ Work mat (Clauss Fiber-Safe™)

➢ Tubing cutter (Ideal 45-162)

➢ Kevlar scissors (with ceramic blade)

➢ Break off blade knife

➢ Cleaver (Alcoa Fujikura CT 04, CT20 or CT30 (recommended), or Unicam cleaver (Figure 8-2 or Figure 8-3)

➢ For training, premises cable, at least 3' long

➢ Unicam connectors with the same core diameter as in the cable:

95-000-50 (ST™-compatible, 62.5-μ, composite ferrule)

95-000-40 (SC, 62.5-μ, composite ferrule)

95-050-51 (ST-™ compatible, 50-μ, zirconia ferrule)

95-050-41 (SC, 50-μ, zirconia ferrule)

95-000-51 (ST-™ compatible, 62.5-μ, zirconia ferrule)

95-000-41 (SC, 62.5-μ, zirconia ferrule)

➢ 99% isopropyl alcohol

➢ Lens grade tissues

➢ Connector installation tool (Figure 8-5)

➢ Fiber collection bottle

Figure 8-5: UniCam Installation Tool

8.3 PROCEDURE

For field installation, the installer repeats this procedure on all connectors on one end of the cable. He repeats this procedure on all connectors on the opposite end.

For training, the installer installs a connector on one buffer tube. He repeats this procedure on the same color buffer tube at the opposite end of the cable.

8.3.1 REMOVE JACKET

For field installation, the installer prepares the cable end to the dimensions specified for the enclosure. For training, the installer prepares the cable ends to expose a minimum buffer tube length of 18".

8.3.2 LOAD TOOL

The installer opens the installation tool by pivoting the top crimping handle so that the handle is off the tool (Figure 8-6). He rotates the cam wrench handle to the vertical position (Figure 8-7).

Figure 8-6: Open UniCam Installation Tool

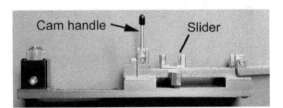

Figure 8-7: Starting Position Of Cam Handle

If necessary, he adjusts the position of the connector cam: it should be at 90° to the date code surface (Figure 8-8)

He removes the cap from the lead-in tube at the back end of the connector (Figure 8-9).

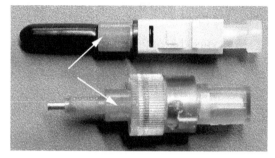

Figure 8-8: Proper Cam Positions

Figure 8-9: Date Code On Top Surface

The installer positions the connector so that the date label (for the SC connector) or 'UP' label (for ST™-compatible connector) is up (Figure 8-9). He pulls the slider towards the right end of the installation tool (Figure 8-7). He places the tip of the connector into the slider.

While guiding the lead-in tube into the cam wrench, he allows the slider to push the connector into and through the cam wrench until the connector is fully seated. The connector is fully seated when the lead-in tube extends beyond the edge of the crimping surface (Figure 8-10 and Figure 8-11).

Figure 8-10: Proper Lead In Tube Position

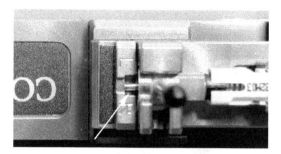

Figure 8-11: Improper Lead In Tube Position

8.3.3 PREPARE FIBER END

The installer installs a boot on the buffer tube. He marks the buffer tube at 40 mm and 52 mm from the end. The 52 mm mark is a process control mark.

By the instructions in 7.3.6.1 or 7.3.6.2, the installer strips the buffer tube and primary coating in increments 1/2" (12.7 mm) or less, to expose 40 mm of bare fiber.

8.3.4 PREPARE CLEAVER

The installer lifts the cleaver outer cover, the inner cover and the breaking arm (Figure 8-12 to Figure 8-14). He moves the scribing arm to the starting position. The starting position is the same side of the cleaver as the outer cover hinge (Figure 8-15). Before the first and after every tenth cleave, he blows out the grooves of the cleaver with lens grade gas.

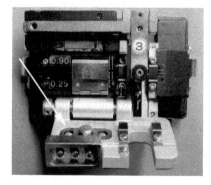

Figure 8-12: Outer Cover Open

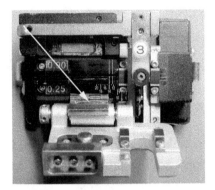

Figure 8-13: Inner Cover Open

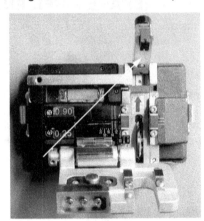

Figure 8-14: Breaking Arm Up

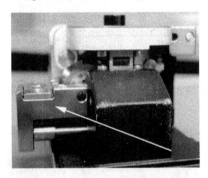

Figure 8-15: Scribing Arm In Starting Position

8.3.5 CLEAN FIBER

The installer moistens a lens grade tissue with isopropyl alcohol. He folds the moist area of the tissue around the fiber. He pulls the fiber through the fold twice. He inspects the fiber to verify lack of contamination of the fiber surface (Figure 7-8).

Caution
Do not place the fiber down.

8.3.6 CLEAVE FIBER

The installer places the buffer tube in the 0.9 groove of the cleaver so that the end of the buffer tube aligns at the 8.5 mm mark. He closes the inner cover (Figure 8-16). If necessary, he readjusts the position of the end of the buffer tube to the 8.5 mm cleave length.

He closes the outer cover (Figure 8-17). He slides the scribing arm in the direction of the arrow on the scribing arm (Figure 8-17 and Figure 8-18).

He lowers the breaking arm, labeled 3. With the tip of his finger (Figure 8-19), he pushes down the black tab of the breaking arm until the outer cover pops up (Figure 8-20).

He holds the buffer tube, lifts the breaking arm, the outer cover, and the inner cover. He places the broken fiber in the fiber collection bottle.

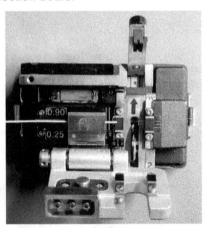

Figure 8-16: Closed Inner Cover

Figure 8-17: Closed Outer Cover

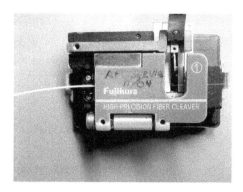

Figure 8-18: Cleaver After Scribing

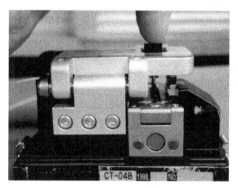

Figure 8-19: Proper Activation Of Breaking Arm

Figure 8-20: Cleaver Cover Popped

Caution: Avoid Contaminating End

Do not place the cleaved fiber onto any surface. Do not touch the sides or the end of cleaved fiber against anything. Do not clean the cleaved fiber. Do not delay the next step.

8.3.7 INSTALL FIBER

While rotating the fiber, the installer inserts the fiber into the lead-in tube until the fiber butts against the internal fiber (Figure 8-21). He checks the process control mark, the 52 mm mark, to ensure that it is within 2 mm of lead-in tube.

If the process control mark is closer than 2 mm to the lead in tube, the fiber has broken. He replaces the connector, restrips and recleaves the fiber.

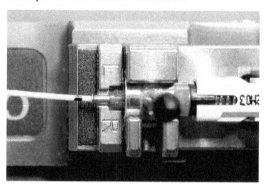

Figure 8-21: Fiber Inserted Into Feed In Tube

The installer inserts the buffer tube into the clip at the end of the installation tool so that the buffer tube bows down by 0.5" from its straight position (Figure 8-22). Gently, he pushes the buffer tube at the lead-in tube to ensure that it is fully inserted. He rotates the cam wrench to the horizontal position (Figure 8-23).

Figure 8-22: Buffer Tube In Rear Clip

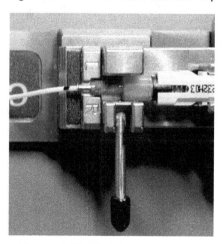

Figure 8-23: Rotated Cam Lever

To crimp the lead-in tube, the installer flips the top crimping handle onto the lead-in tube and presses firmly (Figure 8-24). The lead-in tube will be flat (Figure 8-25).

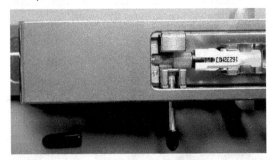

Figure 8-24: Crimping The Lead-In Tube

Figure 8-25: Crimped Lead-In Tube

He flips the crimping handle open. He pulls the slider towards the right end of the installation tool. He removes the connector from the installation tool. He holds the connector and pushes the boot over the lead-in tube and back shell (Figure 8-26).

Note: the installer does not hold the buffer tube and push the boot onto the connector back shell.

Figure 8-26: Boot Installed

8.3.8 VFL EVALUATION

The installer can use a VFL to provide a preliminary indication of the loss of a crimp and cleave connector (Figure 8-27). In addition, this test can indicate which end is high loss. This author's observations are:

➢ No glow in the cam indicates low loss

➢ A faint glow in the cam does not necessarily indicate an unacceptable high loss connector

➢ A bright glow in the cam (a Christmas tree light!) indicates unacceptable high loss (Bottom, Figure 8-27)

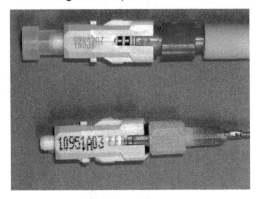

Figure 8-27: VFL Test

This test should be done on an SC connector prior to installation of the outer housing. The installer places a VFL on one end of a fiber. The installer views the connector on the opposite end. If he sees a red glow, light is escaping from the core and the loss of the connector can be high. The installer places a tag on this connector to indicate the potential for high loss.

The VFL is placed on the opposite end because it launches light into the cladding. This cladding light can escape, incorrectly indicating high loss. Even if low loss, cleave and crimp connectors will almost always glow at the end on which the VFL in installed.

8.3.9 SC FINAL ASSEMBLY

The installer positions the date code on the inner housing and the key on the outer housing in the up position (Figure 8-28). He inserts the connector into the outer housing from the end opposite the key until the inner housing snaps through the opposite end (Figure 8-29). He places a cap onto the ferrule end of the connector.

Figure 8-28: Proper Alignment Of Date Code and Key

Figure 8-29: Inner Housing Fully Inserted

8.4 ST-™ COMPATIBLE PROCEDURE

Installation of the ST-™ compatible connector is almost the same as that of the SC connector. The only difference is its orientation in the installation tool. The installer installs the ST™-compatible connector in the tool with the cap on the ferrule. The date code and the word 'UP' on the cap will be up (Figure 8-30).

8.5 TEST LOSS

When he has installed connectors on both ends of a buffer tube or on both ends of all buffer tubes, the installer tests the insertion loss.

Figure 8-30: ST™- Compatible Connector In Tool

8.6 TROUBLESHOOTING

8.6.1 52 MM MARK CLOSER TO LEAD IN TUBE THAN 2 MM

Potential cause: broken fiber due to damage during stripping
Action: replace connector; pull fiber through stripper without bending fiber

8.6.2 HIGH LOSS

Potential cause: contamination of fiber from placing of surface or touching fiber after cleaving
Action: do not place fiber on any surface or touch fiber after cleaving

Potential cause: dirty, worn or damaged scribing blade
Action: clean blade; if problem persists, rotate or replace blade

Potential cause: dirt on end of fiber
Action: install in an environment that does not have floating dust or particles in air

Potential cause: bad cleave due to dirt on fiber
Action: clean fiber prior to cleaving

Potential cause: bad cleave due to dirt on cleaver
Action: clean cleaver pads, scribing blade, and fiber groove

8.7 SUMMARY

Open crimping handle

Rotate cam lever to vertical position

If necessary, adjust cam position of connector

Load connector into tool

Prepare cable end to expose buffer tube

Install boot on buffer tube

Mark buffer tube at 40 mm and 52 mm

Strip the buffer tube to the 40 mm mark

Open cleaver

Clean cleaver before first and after every tenth cleave

Set scribing arm in starting position

Clean the fiber

Position the end of buffer tube at 8.5 mm

Cleave the fiber

Insert the fiber into lead in tube

Verify proper location of second mark

Bow fiber 0.5"

Rotate cam lever to horizontal position

Crimp lead in tube

Remove connector from tool

Slide boot over lead in tube

Optional: VFL test

Install connector into outer housing

Install cap on connector

When connectors are installed on both ends, test cable.

9 REVIEW QUESTIONS

9.1 INSTRUCTIONS

For all questions, assume a preface of 'According to the manual'. Some questions are repeated for each chapter.

9.2 CHAPTER 5

1. List the tools and supplies needed for epoxy installation.

2. Before stripping the buffer tube and primary coating, what should the installer do?

3. What is the maximum recommended strip distance for the buffer tube?

4. What is the prime difference in using a Clauss stripper and a Miller stripper?

5. When stripping a jacketed fiber, what should the installer do with the strength members?

6. When should the installer perform a continuity test of the cable?

7. When does the installer clean the fiber?

8. After cleaning the fiber, for what does the installer inspect the fiber?

9. Describe the strength test.

10. Is the strength test required for every fiber?

11. What are the two purposes of a dry fit?

12. What is purpose of a white light test of the connector?

13. What should the installer do just prior to injecting the connector with epoxy?

14. What can happen if there is insufficient epoxy in the connector?

15. What two effects can occur if there is excess epoxy in the connector?

16. What the are two parameters that determine whether the epoxy cures completely?

17. What prevents the fiber from breaking after scribing?

18. Where should the scriber touch the fiber with the scriber?

19. How should the installer remove the excess fiber?

20. While polishing, what should the installer hold?

21. What is the first polish that the installer performs on a cured connector?

22. Does the first polish require a pad?

23. Prior to polishing, what does the installer clean?

24. Describe the motion of pad polishing.

25. What are the two methods of cleaning of polishing equipment?

26. What is the function of the first pad polish?

27. What is the function of the second pad polish?

28. Is a third pad polish necessary?

29. Prior to changing to a finer film, what should the installer do?

30. What is the sequence of polishing films?

31. After cleaning the ferrule with a wet tissue, what should the installer do?

32. For the SC connector, what must be aligned for final assembly?

33. What is the one significant difference between multimode polishing films and singlemode polishing films?

34. Once the tight tube of a premises cable is exposed, what is the first step the installer should perform?

9.3 CHAPTER 6

1. List the tools and supplies needed for quick cure adhesive installation.

2. Before stripping the buffer tube and primary coating, what should the installer do?

3. What is the maximum recommended strip distance for the buffer tube?

4. What is the prime difference in using a Clauss stripper and a Miller stripper?

5. When stripping a jacketed fiber, what should the installer do with the strength members?

6. When should the installer perform a continuity test of the cable?

7. When does the installer clean the fiber?

8. After cleaning the fiber, for what does the installer inspect the fiber?

9. Describe the strength test.

10. Is the strength test required for every fiber?

11. What are the two purposes of a dry fit?

12. What is purpose of a white light test of the connector?

13. What should the installer do just prior to injecting the connector with adhesive?

14. When should the installer apply the primer?

15. What prevents the fiber from breaking after scribing?

16. Where should the scriber touch the fiber with the scriber?

17. How should the installer remove the excess fiber?

18. What is the first polish that the installer performs on a cured connector?

19. Does the first polish require a pad?

20. While polishing, what should the installer hold?

21. Prior to polishing, what does the installer clean?

22. Describe the motion of pad polishing.

23. What are the two methods of cleaning of polishing equipment?

24. What is the function of the first pad polish?

25. What is the function of the second pad polish?

26. Is a third pad polish necessary?

27. Prior to changing to a finer film, what should the installer do?

28. What is the sequence of polishing films?

29. After cleaning the ferrule with a wet tissue, what should the installer do?

30. For the SC connector, what must be aligned for final assembly?

31. Once the tight tube of a premises cable is exposed, what is the first step the installer should perform?

9.4 CHAPTER 7

1. List the tools and supplies needed for hot melt adhesive installation.

2. Before stripping the buffer tube and primary coating, what should the installer do?

3. What is the maximum recommended strip distance for the buffer tube?

4. What is the prime difference in using a Clauss stripper and a Miller stripper?

5. When stripping a jacketed fiber, what should the installer do with the strength members?

6. When should the installer perform a continuity test of the cable?

7. When does the installer clean the fiber?

8. After cleaning the fiber, for what does the installer inspect the fiber?

9. Describe the strength test.

10. Is the strength test required for every fiber?

11. What are the two purposes of a dry fit?

12. Can the installer perform a white light test of the hot melt connector?

13. What prevents the fiber from breaking after scribing?

14. Where should the scriber touch the fiber with the scriber?

15. How should the installer remove the excess fiber?

16. What is the first polish that the installer performs on a cured connector?

17. Does the first polish require a pad?

18. Prior to polishing, what does the installer clean?

19. While polishing, what should the installer hold?

20. Describe the motion of pad polishing.

21. What are the two methods of cleaning of polishing equipment?

22. What is the function of the first pad polish?

23. Describe the two conditions that can occur at the start of the first pad polish.

24. What is the function of the second pad polish?

25. Prior to changing to a finer film, what should the installer do?

26. What is the sequence of polishing films?

27. After cleaning the ferrule with a wet tissue, what should the installer do?

28. For the SC connector, what must be aligned for final assembly?

29. Once the tight tube of a premises cable is exposed, what is the first step the installer should perform?

9.5   CHAPTER 8

1. List the tools and supplies needed for Unicam® installation.

2. At the start of installation, in what position should the cam arm be?

3. At the start of installation, in what position should the feed in tube be?

4. What is the strip length for a Unicam® connector?

5. What is the cleave length for a Unicam® connector?

6. After cleaving, should the installer clean the fiber?

7. Against what should the installer touch the end of the fiber?

8. After cleaving, where should the installer place the fiber?

9. After insertion, where should the process control mark be?

10. When tested with a VFL, where should a properly installed connector glow?

11. On which end of the cable should the VFL be placed?

12. If a link has high loss and one connector glows, which connector should be replaced?

10 APPENDICES

10.1 REVIEW QUESTION ANSWERS

For most questions, we provide the section(s) and figure(s) in which the answer is. Where necessary, we provide the exact answer.

10.2 CHAPTER 2

1. 2.3.1.1
30. 2.4.5
31. 2.3.1.1
32. 2.3.1.1
33. 2.4.2
34. 2.4.5
35. 2.4.5
36. 2.5
37. 2.5.1
38. 2.5.3
39. 2.5.1.6
40. 2.3.1
41. 2.3
42. 2.3.1.2
43. 2.3.1.2
44. 2.3.10
45. 2.3.10
46. 2.5
47. 2.3.1.3
48. 2.3.1.3
49. 2.5.1.1
50. 2.5.1.8
51. 2.5.1.4
52. 2.5.3.2
53. 2.5.1.1
54. 2.5.1.6
55. 2.5.2.1
56. 2.4.4
57. 2.3.10, 2.3.1.2

10.3 CHAPTER 3

1. 3.4.2, 3.4.3
2. 3.4.3.5
3. 3.3.1.3
4. 3.4.3.5
5. 3.3.2
6. 3.2.1.1.2
7. 3.4.3.3
8. 3.3.2.4
9. 3.3.4.1
10. 3.3.1.3
11. 3.5.2
12. 3.4.3.3
13. 3.4.3
14. Answer

1200/.95=1263 connectors required for high cost cleaver

1200/.87=1379 connectors required for low cost cleaver

(1379-1263) = 116 connectors

116*15= $1740 cost difference

Increased cost of cleaver= $900

High cost cleaver more than pays for itself!

10.4 CHAPTER 4

1. 4.4.1

2. 4.4.2

3. 4.4.2

4. 4.3.3

5. 4.3.3

6. 4.2

7. 4.4.5

8. 4.4.3

9. 4.4.3

10. 4.4.1

11. pistoning, incomplete polishing [Review Question 16]

12. bad; not round; features in core; not flush

13. bad; features in core;

14. bad; not round; features in core

15. bad; not flush

16. bad; feature in core

17. good

18. bad: scratches/feature in core

19. good

20. bad; dirt on core, cladding, and ferrule

21. bad; dirt on ferrule

22. bad; dirt on core, cladding, and ferrule

23. bad; either not flush or cladding not clean

24. good

25. bad: scratch/feature in core

26. bad; feature just barely in core

27. bad; core, cladding, and ferrule not clean

28. bad; core, cladding, and ferrule not clean

10.5 CHAPTER 5

1. 5.2

2. 5.3.6, Figure 5-2

3. 5.3.6

4. 5.3.6.1

5. 5.3.6.2.2

6. 5.3.6.3

7. 5.3.7

8. 5.3.7

9. 5.3.8

10. 5.3.8

11. 5.3.9.2

12. 5.3.9.1

13. 5.3.10

14. hold strength members along the jacket

15. epoxy can 'glue' together parts of the connector that must be free to move; epoxy can wick up the strength member, making the cable stiff

16. 5.3.1, 5.3.13.2

17. 5.3.14

18. 5.3.14

19. 5.3.14

20. 5.3.16.3

21. 5.3.15

22. 5.3.15

23. 5.3.16.1, 5.3.16.2

24. 5.3.16.3

25. 5.3.16.2

26. 5.3.16.3

27. 5.3.16.5

28. 5.3.16.6

29. 5.3.16.4

30. 5.3.16

31. 5.3.17.2

32. 5.3.20

33. 5.3.21.4, 5.3.16.5

34. 5.3.4

10.6   CHAPTER 6

1. 6.2

2. 6.3.5.1

3. 6.3.5.1

4. 6.3.5.1

5. 6.3.5.2.2

6. 6.3.5.3

7. 6.3.6

8. 6.3.6

9. 6.3.7.2

10. 6.3.5.4

11. 6.3.7.2

12. 6.3.7.1

13. 6.3.8

14. 6.3.9

15. 6.3.13

16. 6.3.13

17. 6.3.13

18. 6.3.14

19. 6.3.14

20. 6.3.15.3

21. 6.3.15.2

22. 6.3.15.3

23. 6.3.15.2

24. 6.3.15.3

25. 6.3.15.5

26. 6.3.15.6

27. 6.3.15.

28. 4

29. 6.5.2

30. 6.6, Figure 6-14

31. 6.3.3

10.7   CHAPTER 7

1. 7.2

2. 7.3.6.1

3. 7.3.6.1

4. 7.3.6.1, 7.3.6.2

5. 7.3.4

6. 7.3.6.3 before installing the connector

7. 7.3.7

8. 7.3.7

9. 7.3.6.4

10. 7.3.6.4

11. 7.1

12. 7.3.9

13. 7.3.9

14. 7.3.9

15. 7.3.10

16. 7.3.10

17. 7.3.11.1, 7.3.11.2

18. 7.3.11.3

19. 7.3.11.3

20. 7.3.11.2

21. 7.3.11.3

22. 7.3.11.3

23. 7.3.11.5

24. 7.3.11.4

25. 7.3.11.3, 7.3.11.5

26. 7.6.1.1

27. 5.3.20

28. 7.3.5.1

10.8   CHAPTER 8

1. 8.2

2. 8.3.2

3. 8.3.2, Figure 8-10

4. 8.3.3

5. 8.3.6

6. 8.3.6

7. 8.3.6

8. 8.3.7

9. 8.3.7

10. 8.3.8

11. 8.3.8

12. 8.3.8

10.9 CONNECTORS

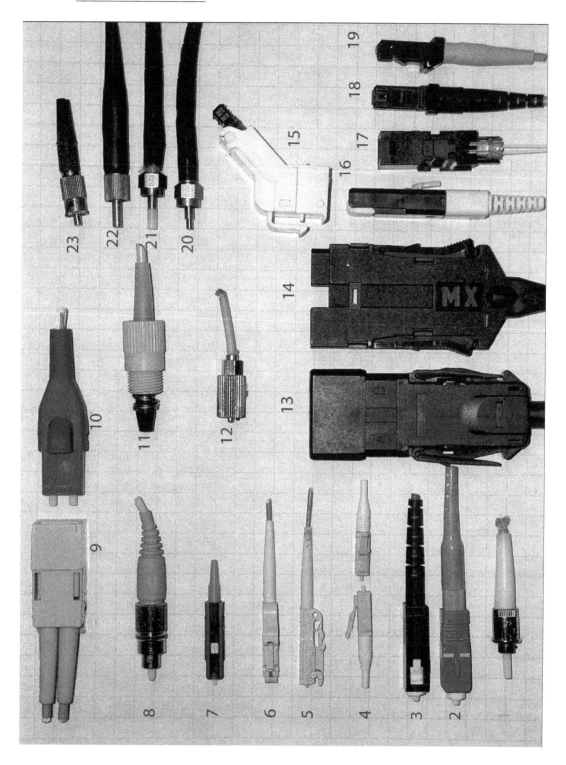

10.10 GLOSSARY

905 SMA: see SMA 905

adapter: a device for mating two connectors. Also known as mating adapter, barrel, bulkhead and feed through

APD: avalanche photodiode. This device converts an optical signal to an electrical signal.

armor: a layer of material, usually stainless steel, which is placed around a cable core to prevent damage from gnawing rodents and from crush loads.

attenuation: the loss of optical power or intensity as light travels through a fiber. It is expressed in units of decibels. When used to describe fibers or cables, it is expressed as a rate in decibels/kilometer. In this book, attenuation refers to reduction in signal strength in the fiber or cable.

back reflection: an outdated term that was used to mean return loss or reflectance. See reflectance and return loss.

back shell: the portion of a connector in back of the retaining nut or latching mechanism.

bandwidth: a measure of the transmission capacity of an analog transmission system.

bandwidth-distance product: the product of the length of a fiber and the analog bandwidth that the fiber can transmit over that length. It is expressed in units of MHz-km for multimode fibers. It is not the relevant parameter for laser-optimized fibers. It is not used for singlemode fibers.

bend radius, long term: see bend radius, minimum unloaded.

bend radius, minimum loaded: the smallest radius to which a cable can be bent during installation at the maximum recommended installation load without any damage to either the fiber or the cable materials. Typically, this radius is 20 times the diameter.

bend radius, minimum unloaded: the smallest radius to which a cable can be bent without any damage to either the fiber or the cable materials while the cable is unloaded. Typically, this radius is 10 times the diameter for the life of the cable.

bend radius, short term: see bend radius, minimum loaded.

biconic: a type of connector.

binding tape: a cable component. This tape holds buffer tubes together during jacket extrusion.

binding yarn: a cable component. This yarn holds buffer tubes together during jacket extrusion.

bit error rate: a measure of the accuracy of a digital fiber optic system. The BER is the rate of errors produced by the optoelectronics. Abbreviated as BER.

bit rate: the data transmission rate of a digital transmission system.

boot: a plastic device that slides over the cable and the back shell. It serves to limit the radius of curvature of the cable as the cable exits the back shell.

break out: a type of cable composed of sub cables, each of which contains a single fiber.

3buffer coating: obsolete term. See 4primary coating.

buffer tube: a layer of plastic that surrounds a fiber or a group of fibers.

bulkhead: see adapter.

butt coupling: a method of transmitting light from one fiber to another by precise mechanical alignment of the two fiber ends without the use of lenses.

BWDP: bandwidth-distance product.

cable: the structure that protects an optical fiber or fibers during installation and use.

cable core: the structure of fibers, buffer tubes, fillers and strength members that reside inside the inner most jacket of a cable.

cable end boxes: the enclosures placed on the end of a cable to protect the buffer tubes and fibers.

cap: a plastic structure that protects the end of a connector ferrule from dust and damage when the connector is not in use.

CBT: see central buffer tube

central buffer tube: a cable design in which all fibers reside in a single, centrally located buffer tube.

central strength member: a strength member that resides in the center of a cable.

chromatic dispersion: the spreading of pulses of light due to rays of different wavelengths traveling at different speeds through the core. See spectral width.

cladding diameter: the outer diameter of the cladding. It is measured in micrometers.

cladding non-circularity: the degree to which the cladding and the core deviate from perfect circularity.

cladding: the region of an optical fiber that confines the light to the core and provides additional strength to the fiber.

cleaver: device to create a flat and perpendicular surface on end of a fiber.

cleaving: the process of creating a fiber end that is flat and perpendicular to the axis of the fiber.

coating: see primary coating

concentricity: the degree to which the core deviates from being in the exact center of the cladding.

cone of acceptance: the cone defined by the critical angle or the numerical aperture. This is the cone within which all of the light enters and exits a fiber.

core diameter: the diameter of the region in which most of the light energy travels. It is measured in micrometers.

core offset: see concentricity.

core: the region of an optical fiber in which most of the light energy travels.

coupler: a device that allows two separate optical signals to be joined for transmission on a single fiber.

crimp ring: the device that is deformed around the back shell of a connector. The crimp ring traps the strength member of the cable, providing acceptable cable-connector strength.

critical angle: the maximum angle to the axis of a fiber at which rays of light will enter a fiber and experience total internal reflection at the core- cladding boundary.

crush load, maximum recommended: the recommended maximum load that can be applied to a fiber optic cable without any permanent change in the attenuation of the cable. This can be specified as a long-term or a short-term crush load.

cut-off wavelength: the wavelength, below which a singlemode fiber will not transmit a single mode. Below this wavelength, the singlemode fiber will transmit multimode manner.

D4: a connector type

design: used in this text to refer to a cable

dielectric: having no components that conduct electricity in the cable.

differential modal attenuation (DMA): the mechanism by which rays in multimode fibers experience differing attenuation rates due to their mode, or location in the core.

differential modal delay (DMD): a measurement of dispersion in a multimode fiber; DMD is the measurement used for bandwidth measurement of laser-optimized for fiber.

dispersion: the spreading of pulses in fibers.

ESCON™: a connector type.

expanded beam coupling: a method of transmitting light from one fiber to another with lenses.

FC: a connector type.

FC/PC: a connector type.

FDDI: fiber data distributed data interface.

feed through: see adapter.

ferrule: that portion of the connector that aligns a fiber. A ferrule is present in all connector types except the Volition™.

fiber: the structure that guides light in a fiber optic system.

Fiber Jack: see Opti-Jack™.

filled and blocked cable : a type of cable in which all empty space is filled with compounds to prevent moisture ingress. Filling refers to a gel material in buffer tubes. Blocked refers to grease outside of buffer tubes.

fillers: cable materials that fill otherwise empty space in a cable.

Fresnel reflection: the reflection that occurs when light travels between two media in which the speed of light differs.

FRP strength member: a fiberglass reinforced plastic or epoxy rod that is used as a dielectric strength member in cables. The term 'plastic' may refer to either a polymer plastic or an epoxy.

fusion splice: a splice made by melting two fibers together.

gel-filling compound: a compound placed inside a loose buffer tube to prevent water from contacting the fiber(s) in that buffer tube.

graded index: a type of multimode fiber, in which the chemical composition of the core is not uniform.

HCS™: a hard clad silica fiber.

heat shrink tubing: tubing placed on the back shell of a connector.

HiPPI: high speed, parallel processor interface.

index of refraction (IR, h): the ratio of the speed of light in the material to the speed of light in a vacuum. This is a dimensionless number.

inner duct: a corrugated plastic pipe in which fiber optic cables are placed.

inner jacket: any layer of jacketing plastic that is not the outer-most layer.

installation strength, maximum recommended: the maximum load that can be applied along the axis of a cable without any damage to the fibers .

installation temperature range: the temperature range within which a cable can be installed without damage.

jacket: a layer of plastic in a cable. It can be an outer jacket or an inner jacket.

jumper: a short length of a single fiber cable with connectors on both ends.

Kevlar®: an aramid yard produced by DuPont Chemical. It is used to provide strength in fiber optic cables.

keying: a connector mechanism by which ferrules are prevented from rotating.

laser diode: a semiconductor that converts electrical signals to optical signals.

laser optimized (LO) a multimode fiber that is designed to enable long transmission distance at 1 Gbps-100 Gbps.

latching mechanism: device that retains a connector to a receptacle or an adapter .

LC: a SFF connector with a 1.25 mm ferrule.

LX.5: a SFF connector with a 1.25 mm ferrule and a built-in dust cover.

LED: light emitting diode. A semiconductor that converts electrical signals to optical signals.

lensed coupling: see expanded beam coupling.

loose buffer tube: a buffer tube with space between the outer diameter of the primary coating and the inner diameter of the buffer tube.

loose tube: a cable design in which the fiber floats loosely inside an oversized tube.

loss: the end-to-end reduction in optical power as light travels through a fiber, connectors or splices. In this book, the 'loss' refers to reduction of optical power at a splice or a connector pair.

material dispersion: the spreading of pulses of light due to rays of light traveling through different regions of the core.

maximum recommended installation load: see installation strength, maximum recommended.

mechanical splice: a mechanism that aligns two fiber ends precisely for efficient transfer of light from one fiber to another.

MFPT: a multiple fiber per (buffer) tube cable design. This design usually has 6 or 12 fibers per loose buffer tube.

mini-BNC: a connector type.

minimum recommended long term bend radius: see bend radius, minimum unloaded.

minimum recommended short-term bend radius: see bend radius, minimum loaded.

MPO: the generic term for a 4-24 fiber, connector with a single ferrule. The usual form of use is MPO/MPT.

MPT: the proprietary term for a 4-12 fiber, connector with a single ferrule. The usual form of use is MPO/MPT.

modal dispersion: the spreading of pulses of light due to rays traveling different paths through a multimode fiber.

modal pulse spreading: see modal dispersion.

mode: a path in which light can travel in a fiber core. Mode translates roughly to 'path.'

mode field diameter: the diameter within which the light energy field travels in a singlemode fiber.

monomode: see singlemode. This term is used outside of North America.

MT-RJ: a duplex SFF connector with a single ferrule.

MU: a SFF connector with a 1.25 mm ferrule and a size of approximately half that of the SC connector.

multimode: a method of propagation of light in which all of the rays of light do not travel in a path parallel to the axis of the fiber.

NA: see numerical aperture.

numerical aperture (NA): the sine of the critical angle. The NA is a measure of the solid angle within which rays of light will enter and exit the fiber.

offset, core: see core offset.

optical amplifier: a device that increases the signal strength without an optical to electrical to optical conversion.

optical coupling: see expanded beam coupling.

optical power budget: the maximum loss of optical power which transmitter-receiver pair can withstand while still functioning at the specified level of accuracy.

Opti-Jack™: a duplex SFF connector.

optical return loss: see return loss.

optical rotary joint: a rotating joint that allows transmission of light from a stationary fiber to a rotating fiber.

optical switch: a switch that can direct light to more than one output path.

optical time domain reflectometer (OTDR): a test device that creates a map of the loss of signal strength of an optical path.

optical waveguide: another term for an optical fiber.

optoelectronic device: any device that converts a signal from electrical to optical domain or vice versa.

optoelectronics: see optoelectronic device.

OTDR: optical time domain reflectometer or optical time domain reflectometry.

ovality: a measure of the degree to which a fiber deviates from perfect circularity. See cladding non-circularity.

passive component: a device that manipulates light without requiring an optical to electrical signal conversion.

patch panel: a sheet of material that contains adapter(s).

PCS: a plastic clad silica fiber.

PD: photodiode. This device converts an optical signal to an electrical signal.

pigtail: a length of fiber or cable that is permanently attached to a connector or an optoelectronic device.

Ping-Pong: a type of transmission that results from using a single LED as both a transmitter and receiver.

plug: another term for connector.

POF: plastic optical fiber. A fiber with a plastic core and a plastic cladding.

polishing fixture: device to hold connector perpendicular to fiber end;

polishing puck: see polishing fixture

polishing tool: see polishing fixture

primary coating: a layer of plastic placed around the cladding by the fiber manufacturer.

primary coating diameter: the diameter of the layer of plastic that is placed around the fiber by the fiber manufacturer.

pull-proof: a performance characteristic of connector types. A connector type is pull-proof when tension on the cable attached to the connector does not produce an increase in the loss of the connector.

pulse dispersion: see dispersion.

pulse spreading: see dispersion.

receptacle: the device within which an active device is mounted. The receptacle is designed to mate with a specific connector type.

reflectance: a measure of the ratio of reflected power to incident power for a single device, such as a connector or mechanical splice. Reflectance is measured in units of dB.

reinforced jacket: two layers of plastic that are separated by strength members.

repeatability: the maximum change in loss between successive measurements of the loss of a connector pair.

retaining nut: device that attaches a connector to receptacle or to an adapter

return loss: the ratio of incident power to power reflected or back scattered for an entire.

ribbon: a structure on which multiple fibers are precisely aligned.

SAP: see super absorbent polymer

super absorbent polymer: a material that absorbs moisture by converting it to a gel; used in fiber optic cables to provide moisture resistance; incorporated into cables as tapes, yarns, threads and powders.

SC: a connector type.

SFF: see 'small form factor'.

shrink tubing: plastic that covers back shell of connector

simplex: a single fiber cable.

singlemode: a method of propagation of light in which all of the energy of light travels the same path length

slotted core: a cable design with a core containing helical slots.

SMA 905: a connector type

SMA 906: a connector type

small form factor: a type of connector with the characteristic of small size that enables doubling of density in patch panels, enclosures, and switches.

spectral width: the measure of the width of the output power-wavelength curve at a power level equal to half the peak power.

splice enclosure: a structure that encloses and protects splice trays and cable ends.

splice tray: a structure that encloses and protects fibers.

splice: a device for permanent alignment of two fiber ends.

splitter: a device that creates multiple optical signals from a single optical signal.

spot size: the size of the area of an LED or laser diode from which light is produced.

ST®: the first of a series of connector types designed by ATT. Other types from ATT are the ST-II and the ST-II+.

star core: see slotted core.

ST-compatible: a connector with a type which is compatible with the ST® connector.

step index: a type of multimode fiber, in which the chemical composition of the core is uniform.

storage temperature range: the temperature range within which a cable can be stored without damage.

strength members: those elements of a cable design that provide strength.

type: the sum of characteristics that differentiates one connector type from another type.

TECS™: a technically enhanced clad fiber similar to hard clad silica fiber.

temperature operating range: the range of temperature within which the cable can be operated during its lifetime without degradation of its properties.

tight tube: a design in which the tube does contact the entire circumference of the fiber. A tight tube can contain only 1 fiber.

total internal reflection: the mechanism by which optical fibers confine light to the core.

type: used in this text to refer to a fiber or connector

use load, maximum recommended: the maximum longitudinal load that can be applied to a cable during its entire lifetime without damage.

VCSEL: vertical cavity, surface emitting laser. A type of relatively low cost multimode light source that transmits at 1-10 Gbps.

Volition™: a duplex SFF connector with no ferrules.

Water-blocking compound: a compound placed in the interstices between buffer tubes or between jackets in a cable.

wavelength division multiplexer: a passive device that combines optical signals with different wavelengths on different fibers light onto a single fiber. The wavelength division demultiplexer performs the reverse function.

wavelength: a measure of the color of the light. It is a length stated in nanometers (nm) or in micrometers (μ).

wiggle proof: a connector performance characteristic. A connector type is wiggle proof if lateral pressure on the back shell does not produce an increase in the loss.

window: the wavelength range within which a fiber is designed to provide specified performance.

Zip cord®: a two-fiber cable with a figure 8 cross-section that allows each of the two fibers to be separated in the same manner as a lamp cord.

10.11 ACRONYMS

ADSS	all dielectric self support
APC	angled physical contact connector
ATM	Asynchronous Transfer Mode
BER	bit error rate
CATV	Cable TV
CWDM	coarse wavelength division multiplexing
D4	a connector
DAC	dual attachment concentrator
DAS	dual attachment station
DS	dispersion shifted
DS-NZD	dispersion shifted, non-zero dispersion
DWDM	dense wavelength division Multiplexing
EDFA	erbium doped fiber amplifier
ESCON	Enterprise System Connection
FC	a connector type
FDDI	fiber data distributed interface
FOLS	Fiber Optic LAN Section of the TIA
FTTD	fiber to the desk
FTTH	fiber to the home
FTTP	fiber to the premises
FTTX	any of the above three
GI	graded index
HDPE	high-density polyethylene

IP	Internet Protocol
LAN	local area network
LD	laser diode
LEAF™ (Corning Inc.)	large effective area fiber
LED	light emitting diode
LSA	least squares analysis
LX.5	a connector type
MFPT	multiple fiber per tube
MIC	media interface connector
MM	multimode
MT-RJ	a duplex connector type
MU	a SFF connector type
NA	numerical aperture
NDS	non-dispersion shifted
NEC	National Electrical Code
NIST	National Institute of Science and Technology
OFC	optical fiber cable, conductive, horizontal rated
OFCP	optical fiber cable, conductive, plenum rated
OFCR	optical fiber cable, conductive, riser rated
OFN	optical fiber cable, non-conductive, horizontal rated
OFNP	optical fiber cable, non-conductive, plenum rated
OFNR	optical fiber cable, non--conductive, riser rated
OPBA	optical power budget available
OPBR	optical power budget requirement
PC	physical contact
PMD	polarization mode dispersion
PON	passive optical network
SAN	storage area network
SAP	super absorbent polymer
SAC	single attachment concentrator
SAS	single attachment station
SC	a connector type
SDH	Synchronous Digital Hierarchy
SI	step index
SM	singlemode
SONET	synchronous optical network
STP	shield twisted pair
UPC	ultra physical contact
UTP	unshielded twisted pair
VCSEL	vertical cavity, surface emitting laser
WDM	wavelength division multiplexing
ZDW	zero dispersion wavelength

10.12 THE AUTHOR

For the last 34 years Mr. Eric R. Pearson has been worked in fiber optic communications. This involvement includes a wide variety of activities, as detailed below.

Mr. Pearson has developed and run two fiber optic cable manufacturing facilities and organizations (Manufacturing Manager, Times Fiber Communications and Business Manager, Whitmor Waveguides). In these positions, Mr. Pearson developed cable designs, manufacturing techniques and qualified designs against performance specifications. As business manager, he was responsible for developing a profitable multi-million dollar business unit.

Mr. Pearson managed two fiber-manufacturing facilities, one for Corning Glass Works and for Times Fiber Communications.

Mr. Pearson has delivered 507 training presentations and trained more than 8400 personnel in proper installation and design procedures. Between his field installations and training, he has made and supervised more than 46,648 connectors. Mr. Pearson has performed tens of thousands of OTDR, ORL, insertion loss, and dispersion tests on both multimode and singlemode cables.

From both field experience and training, Mr. Pearson gained considerable experience to write three definitive texts on cable and connector installation, The Complete Guide to Fiber Optic Cable Installation (Delmar Publishers, 1997, ISBN #0-8273-7318-X), Successful Fiber Optic Installation- The Essentials (2005), Professional Fiber Optic Installation- The Essentials For Success (2011, ISBN 9780976975434) and this text.

He has written the books: Fiber Optic Network Design, Practical Fiber Optic System Design and Implementation, How to Specify and Choose Fiber Optic Cables, and How to Specify and Choose Fiber Optic Connectors.

Mr. Pearson has been a technical expert for a patent infringement suits, law suits between operators and end users, law suits regarding technical fraud in fiber optic cables, and performance specifications.

From 1995-1997, Mr. Pearson was a Director and the Director of Certification, of the Fiber Optic Association (FOA). As the latter, he is responsible for developing requirements and examinations for basic to advanced certification of fiber optic operator personnel. These activities require an in-depth knowledge of all aspects of cable and connector installation.

From the Fiber Optic Association, he has received four advanced Certified Fiber Optic Specialist certifications (CFOS/T, CFOS/S, CFOS/C, and CFOS/I). In 2011, FOA has designated Mr. Pearson as a 'Master Instructor'.

Since 1999, Mr. Pearson has been a Master Instructor for the Building Industry Consultants Services International (BICSI), and the developer of the BICSI fiber optic network design program.

From 1986-2004, he was a Member, Editorial Advisory Board, Fiberoptic Product News.

The Academy of Professional Consultants & Advisors (APCA) certified him as a Certified Professional Consultant (CPC).

He has over 100 articles, reports and presentations to his credit. He is frequently quoted in fiber optic and related trade journals.

He has been selected to speak at three Newport Fiber Optic Marketing Conferences.

He is listed in: Who's Who Worldwide, Who's Who of Business Leaders, Who's Who in Technology and Who's Who in California.

Mr. Pearson has provided consulting services to hundreds of companies in the areas of fiber network design and specification, technical and marketing evaluations.

Mr. Pearson received his education at Massachusetts Institute of Technology (BS, 1969) and Case-Western Reserve

University (MS, 1971). Both degrees are in Metallurgy and Materials Science.

Mr. Eric R. Pearson, CFOS

10.13 SERVICES

Training Programs
(http://www.ptnowire.com/training-list.htm)

Basic Installation (4 days)
http://www.ptnowire.com/Fpro1.htm

Basic Installation With FOA CFOT
Certification (5 days)

Basic With Advanced Connector
Installation (5 days)
http://www.ptnowire.com/Fpro1-2.htm

Advanced Connector Installation (2 days)
http://www.ptnowire.com/Fpro2.htm

Advanced Testing (3 days)

Advanced Splicing (5 days)

Advanced With FOA CFOS Certification

Sales Training

Books

Professional Fiber Optic Installation,
The Essentials For Success (ISBN
9780976975434, 2011)

Mastering The OTDR- Trace
Acquisition And Interpretation (ISBN
9780976975434, 2011)

Mastering Fiber Optic Connector
Installation-The Essentials For
Professional Success (2011)

Mastering Fiber Optic Testing (in
development)

Fiber Optic Network Design (in
development)

Consulting
(http://www.ptnowire.com/services.htm)

Network Design Review

Lawsuit Technical Support

Technical Support For Marketing

Used Splicing And Test Equipment

INDEX